Current Progress in Automation

Current Progress in Automation

Edited by **Neil Green**

CLANRYE
INTERNATIONAL

New Jersey

Published by Clanrye International,
55 Van Reypen Street,
Jersey City, NJ 07306, USA
www.clanryeinternational.com

Current Progress in Automation
Edited by Neil Green

International Standard Book Number: 978-1-63240-127-4 (Hardback)

Printed in the United States of America.

Contents

Preface

This book aims to highlight the current researches and provides a platform to further the scope of innovations in this area. This book is a product of the combined efforts of many researchers and scientists, after going through thorough studies and analysis from different parts of the world. The objective of this book is to provide the readers with the latest information of the field.

The role of Automation in the modern world according to Dr. Kongoli is firmly linked with the modern requirement for sustainable development in the 21st century. One of the fundamentals of sustainable development is "Doing More with Less" which is also one of the aims of automation by substituting human labor with the use of machines. Automation not only escalates productivity and quality but also saves time and energy for humans to deal with the new challenges. This book covers significant topics such as automation in aviation, power system and substation automation, and signal processing in precision dimension measurement.

I would like to express my sincere thanks to the authors for their dedicated efforts in the completion of this book. I acknowledge the efforts of the publisher for providing constant support. Lastly, I would like to thank my family for their support in all academic endeavors.

Editor

Optimization of IPV$_6$ over 802.16e WiMAX Network Using Policy Based Routing Protocol

David Oluwashola Adeniji
University of Ibadan
Nigeria

1. Introduction

Internet application needs to know the IP address and port number of the remote entity with which it is communicating during mobility. Route optimization requires traffic to be tunneled between the correspondent node(CN) and the mobile node (MN).). Mobile IPv6 avoids so-called triangular routing of packets from a correspondent node to the mobile node via the Home Agent. Correspondent nodes now can communicate with the mobile node without using tunnel at the Home Agent..The fundamental focal point for Optimization of IPv6 over WiMAX Network using Policy Based Routing Protocol centered on the special features that describe the goal of optimization mechanism during mobility management. To reiterate these features there is need to address the basic concept of optimization as related to mobility in mobile IPv6 Network. WiMAX which stands for Worldwide Interoperability for Microwave Access, is an open, worldwide broadband telecommunications standard for both fixed and mobile deployments. The primary purpose of WiMAX is to ensure the delivery of wireless data at multi-megabit rates over long distances in multiple ways. Although WiMAX allows connecting to internet without using physical elements such as router, hub, or switch. It operates at higher speeds, over greater distances, and for a greater number of people compared to the services of 802.11(WiFi).A WiMAX system has two units. They are WiMAX Transmitter Tower and WiMAX Receiver. A Base Station with WiMAX transmitter responsible for communicating on a point to multi-point basis with subscriber stations is mounted on a building. Its tower can cover up to 3,000 Sq. miles and connect to internet. A second Tower or Backhaul can also be connected using a line of sight, microwave link. The Receiver and antenna can be built into Laptop for wireless access.This statement brings to the fact that if receiver and antenna are built into the laptop, optimization can take place using a routing protocol that can interface mobile IPv6 network over WiMAX 802.16e.However the generic overview of optimization possibilities most especially for a managed system can be considered.

What then is Optimization? Basically optimization is the route update signaling of information in the IP headers of data packets which enable packets to follow the optimal path and reach their destination intact. The generic consideration in designing route optimization scheme is to use minimumsignaling information in the packet header. Furthermore the delivery of managed system for optimization describes the route optimization operation and the mechanism used for the optimization. In order for optimization to take place, a protocol called route optimization protocol must be introduced. Route optimization protocol is used basically to improve performance. Also route optimization is a technique that enables mobile node and a correspondent node to communicate directly, bypassing the home agent completely; this is based on IPv6 concept.

The use of domains enables a consistent state of deployment to be maintained. The main benefits of using policy are to improve scalability and flexibility for the management system. Scalability is improved by uniformly applying the same policy to large sets of devices , while flexibility is achieved by separating the policy from the implementation of the managed system. Policy can be changed dynamically, thus changing the behavior and strategy of a system, without modifying its implementation or interrupting its operation. Policy-based management is largely supported by standards organizations such as the Internet Engineering Task Force (IETF) and the Distributed Management Task Force (DMTF) and most network equipment vendors.However the Architectures for enforcing policies are moving towards strongly distributed paradigms, using technologies such as mobile code, distributed objects, intelligent agents or programmable networks.

Mobile IP is the standard for mobility management in IP networks. New applications and protocols will be created and Mobile IP is important for this development. Mobile IP support is needed to allow mobile hosts to move between networks with maintained connectivity. However internet service driven network is a new approach to the provision of network computing that concentrates on the services you want to provide. These services range from the low-level services that manage relationships between networked devices to the value-added services provided to the end-users. The complexity of the managed systems results in high administrative costs and long deployment cycles for business initiatives, and imposes basic requirements on their management systems. Although these requirements have long been recognized, their importance is now becoming increasingly critical. The requirements for management systems have been identified and can be facilitated with policy-based management approach where the support for distribution, automation and dynamic adaptation of the behaviour of the managed system is achieved by using policies. IPV6 is one of the useful delivery protocols for future fixed and wireless/mobile network environment while multihoming is the tools for delivering such protocol to the end users. Optimization of Network must be able to address specific market requirements, deployment geographical, end-user demands, and planned service offerings both for today as well as for tomorrow.

2. IP mobility

The common mechanism that can manage the mobility of all mobile nodes in all types of wireless networks is the main essential requirement for realizing the future ubiquitous computing systems. Mobile IP protocol V_4 or V_6 considered to be universal solutions for mobility management because they can hide the heterogeneity in the link-specific technology used in different network. Internet application needs to know the IP address and port number of the remote entity with which it is communicating during mobility. From a network layer perspective, a user is not mobile if the same link is used, regardless of location. If a mobile node can maintain its IP address while moving, it makes the movement transparent to the application, and then mobility becomes invisible. From this problem the basic requirement for a mobile host is the

Mobile IP works in the global internet when the mobile node (MN) which belong to the home agent (HA) moves to a new segment, which is called a foreign network (FN). The MN registers with the foreign agent (FA) in FN to obtain a temporary address i.e a care of address(COA).The MN updates the COA with the HA in its home network by sending BU update message. Any packets from the corresponding node (CN) to MN home address are intercepted by HA. HA then use the BU directly to the CN by looking at the source address of the packet header.

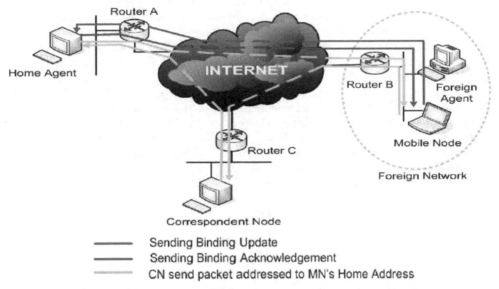

Fig. 1. Mobile IP Network

The most critical challenges of providing mobility at the IP layer is to route packets efficiently and securely. In the mobile IP protocol all packets are sent to a mobile node while away from home are intercepted by its home agent and tunneled to the mobile node using IP encapsulation within IP.

2.1 Limitation of mobile IP

Mobile IP can only provide continuous Internet connectivity and optimal routing to a mobile host, and are not suitable to support a mobile network. The reasons is that, not all devices in a mobile network is sophisticated enough to run these complicated protocols.Secondly, once a device has joined a mobile network, it may not see any link-level handoffs even as the network moves.

2.2 Detailed description of NEMO

Network Mobility describes the situation of a router connecting an entire network to the Internet dynamically changes its point of attachment. The connections of the nodes inside the network to the Internet are also influenced by this movement. A mobile network can be connected to the Internet through one or more MR, (the gateway of the mobile network), there are a number of nodes (Mobile Network Nodes, MNN) attached behind the MR(s). A mobile network can be local fixed node, visiting or nested. In the case of local fixed node: nodes which belong to the mobile network and cannot move with respect to MR. This node are not able to achieve global connectivity without the support of MR. Visiting node belong to the mobile network and can move with respect to the MR(s).Nested mobile network allow another MR to attached to its mobile network. However the operation of each MR remains the same whether the MR attaches to MR or fixed to an access router on the internet. Furthermore in the case of nested mobile network the level mobility is unlimited, management might become very complicated. In NEMO basic support it is important to note that there are some mechanism that allow to allow mobile network nodes to remain connected to the Internet and continuously reachable at all times while the mobile network they are attached to changes its point of attachment.Meanwhile, it would also be meaningful to investigate the effects of Network Mobility on various aspects of internet communication such as routing protocol changes, implications of real-time traffic, fast handover and optimization. When a MR and its mobile network move to a foreign domain, the MR would register its care-of-address (CoA) with it's HA for both MNNs and itself. An IP-in-IP tunnel is then set up between the MR and it's HA. All the nodes behind the MR would not see the movement, thus they would not have any CoA, removing the need for them to register anything at the HA. All the traffic would pass the tunnel connecting the MR and the HA. Figure 2.3 describes how NEMO works.

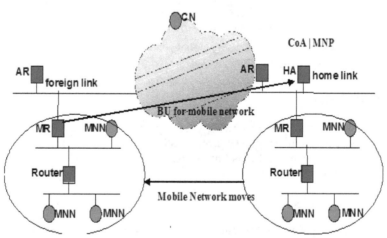

Fig. 2. Network Mobility

2.3 Micro mobility and Macro mobility

This discussion on Micromobility and Macromobility is centered on wireless communication architecture that focuses on the designed of IP Micromobility protocol that compliment an IETF standard for Macromobility management which is usually called Mobile IP. From this point of view, Macro-mobility concerns with the management of users movements at a large scale, between different wide wirelesses accesses networks connected to the Internet. Macro-mobility is often assumed to be managed through Mobile IP. On the other hand, Micro-mobility covers the management of users movements at a local level, inside a particular wireless network.The standard Internet Protocol assumes that an IP address always identifies the node's location in the Internet. This means that if a node moves to another location in the Internet, it has to change its IP address or otherwise the IP packets cannot be routed to its new location anymore. Because of this the upper layer protocol connections have to be reopened in the mobile node's new location.

The Technologies which can be Hierarchical Mobile IP,Cellular IP,HAWAII at different micro mobility solutions could coexist simultaneously in different parts of the Internet. Even at that the Message exchanges are asymmetric on Mobile IP. In cellular networks, message exchanges are symmetric in that the routes to send and receive messages are the same. Considering the mobile Equipment the appropriate location update and registration separate the global mobility from the local mobility. Hence Location information is maintained by routing cache. During Routing most especially in macro mobility scenario a node uses a gateway discovery protocol to find neighboring gateways Based on this information a node decides which gateway to use for relaying packets to the Internet. Then, packets are sent to the chosen gateway by means of unicast. With anycast routing, a node leaves the choice of gateway to the routing protocol.

The routing protocol then routes the packets in an anycast manner to one of the gateways.In the first case, a node knows which gateway it relays its packets to and thus is aware of its macro mobility.

The comparative investigation of different requirement between Micro mobility and Macro mobility are based on following below:

- *Handoff Mobility Management Parameter.*The interactions with the radio layer, initiator of the handover management mechanism, use of traffic bicasting were necessary. The handoff latency is the parameter time needed to complete the handoff inside the network. Also potential packet losses were the amount of lost packets during the handoff must be deduced. Furthermore the involved stations: the number of MAs that must update the respective routing data or process messages during the handover are required.
- *Passive Connectivity* with respect to paging required an architecture that can support via paging order to control traffic against network burden.This architecture is used to support only incoming data packet.Therefore the ratio of incoming and outcoming communication or number of handover experienced by the mobiles are considered for efficiency support purposes.To evaluate this architecture an algorithm is used to perform the paging with respect to efficiency and network load.
- *Intra Network Traffic* basic in micro mobility scenario are the exchanged of packet between MNs connected to the same wireless network. This kind of communication is a large part of today's GSM communications and we can expect that it will remain an important class of traffic in future wireless networks.
- **Scalability and Robustness**: The expectation is that future large wireless access networks will have the same constrains in terms of users load. These facts are to be related to the increasing load of today's Internet routers: routing tables containing a few hundreds of thousandsentries have become a performance and optimization problem.

The review of micromobility via macromobility of key management in 5G technology must addressed the following features:

- 5G technology offer transporter class gateway with unparalleled consistency.
- The 5G technology also support virtual private network
- Remote management offered by 5G technology a user can get better and fast solution.
- The high quality services of 5G technology based on Policy to avoid error.

5G technology offer high resolution for crazy cell phone user and bi-directional large bandwidth shaping.

2.4 Concept of Multihoming

Mobile networks can have multiple points of attachment to the internet, in this case they are said to be multihomed. Multihoming arises when the MR has multiple addresses, multiple egress interfaces on the same link, or multiple egress interfaces on different links. Basically the classification of configuration can be divided into :*Configuration-Oriented Approach, Ownership-oriented Approach* and *Problem-Oriented Approach.* The multihoming analysis classifies all these configuration of multihomed mobile networks using (x, y, z) notation. Variables x, y, and z respectively mean the number of MRs connected to the Internet (so called root MRs), the number of HAs, and the number of Mobile Network Prefix (MNP) s. In case of 1, each variable implies that there exists a single node or prefix. If the variable is N, then it means that one or more agents or prefixes exist in a single mobile network. From different combinations of the 3-tuple (x, y, z), various types of multihoming scenarios are possible. For example the (N, 1, 1) scenario means there is multiple MRs at the mobile network, but all of MRs are managed by single HA and use same MNP.

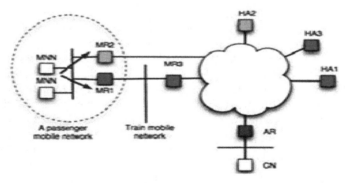

Fig. 3. A Multihoming of Nested Mobile Network

The Figure above shows how a train provide a Wifi network to the passengers with MR3, the passenger could connect to MR3 with MR1 (for example his Laptop). The passenger could also connect directly to Internet with MR2 (his Phone with its GPRS connectivity).The train is connected to Internet with Wimax connectivity. The MNNs can be a PDA and some sensors.

Multihoming from the above nested mobile network provides the advantages of session preservation and load sharing. During optimization the key data communication that must be taking into consideration are:

- Session preservation by redundancy.-the session must be preserved based on the available stable mobile environment either via wireless or wired.

- Load balancing by selecting the best available interface or enabling multiple interfaces simultanousely.Traffic load balancing at the MR is critical since in mobile networks, all traffic goes through the MR.

The apprehension from the above can be justified by specifying the mobile message notification mobile node as well as the procedure for node joining. A mobility notification message contains two important information:(i) the notification interval for multihoming; and (ii)the prefix of the access network that the sending gateway belongs. The optimal choice of the notification interval depends on the mobility of the nodes as well on the amount of traffic sent.

Processing of packets from multihomed nodes is more complex and requires the gateway to perform two tasks. First,the gateway has to verify if a node has recently been informed that its packets are relayed through this access network. If this does not take place, the gateway sends a mobility notification message to the mobile node to inform it about the actual access network. For reducing the amount of mobility notification messages, the gateway records the node address combined with a time stamp in a lookup table. After a notification interval, the gateway deletes the entry and if it is still relaying packets for this node, notifies the mobile node again.Secondly the gateway substitutes the link-local address prefix of the IP source address of the packet with the prefix of the access network it belongs to and forwards the packet to the Internet. When a multihoming node receives a mobility notification message, it adjusts its address prefix to topologically fit the new access network. Subsequently,it informs about its address change using its IP mobility management protocols. In the case where packets of a node are continuously forwarded over different access networks,multihoming support is an advantage to prevent continuous address changes. When a multihoming node receives a mobility notification message, it checks if it already is aware of X or Y access network.

2.5 Requirement of Multihoming configuration

The requirement for Multihomed configurations can be classified depending on how many MRs are present, how many egress interfaces, Care-of Address (CoA), and Home Addresses (HoA) the MRs have, how many prefixes (MNPs) are available to the mobile network nodes, etc. The reader of this chapter should note that there are eight cases of configuration of multihomed mobile network. The 3 key parameter associated to differentiate the configuration are referred to 3-tuple X, Y, Z. To describe any of this requirement configurationin respect to macro mobility, a detection mechanism and notification protocol is required. The below table present the most significant features of the eight classification approach for NEMO. Although there are several configuration but NEMO does not specify any particular mechanism to manage multihoming.

	Configuration	Class	Requirement	Prefix Advertisement
1	Configuration 1, 1, 1	Class 1,1,1	MR, HA, MNP	1 MNP
2	Configuration 1, 1, 1	Class 1,1,n	1 MR, 1 HA, More MNP	2 MNP
3	Configuration 1, 1, 1	Class 1,n,1	1 MR, More HA, 1 MNP	1 MNP
4	Configuration 1, 1, 1	Class 1,n,n	1 MR, More HA, More MNP	Multiple MNP
5	Configuration 1, 1, 1	Class n,1,1	More MR, 1 HA, 1 MNP	MNP
6	Configuration 1, 1, 1	Class n,1,n	More MR, 1 HA, More MNP	Multiple MNP
7	Configuration 1, 1, 1	Class n,n,1	More MR, More HA, MNP	1 MNP
8	Configuration 1, 1, 1	Class n,n,n	More MR, More HA, More MNP	Multiple MNP

Table 1. Analysis of Eight cases of multihoming configuration

What then about the reliability of these configurations during NEMO? Internet connection through another interface must be reliable. The levels of *redundancy* cases can be divided into two: if the mobile node's IP address is not valid any more, and the solution is to use another available IP address; the order is that the connection through one interface is broken, and the solution is to use another. If one of the interfaces is broken then the solution is to use another interface using a *Path Exploration*. However in the case of multiple HAs, the redundancy of the HA is provided, if one HA is broken, another one could be used. The important note here is the broken of one interface can also lead to failure. *Failure Detection* in all the cases in which the number of MNP is larger than 1, because the MNN could choose its own source address, if the tunnel to one MNP is broken, related MNNs have to use another source address which is created from another MNP. In order to keep sessions alive, both failure detection andredirection of communication mechanisms are needed. If those mechanisms could not perform very well, the transparent redundancy can not be provided as well as in the cases where only one MNP is advertised.

The only difference between using one MR with multiple egress interfaces and using multiple MRs each of which only one egress interface is *Load sharing*. Multiple MRs could share the processing task comparing with only one MR, and of course it provides the redundancy of the disrupting of the MR. Therefore the mechanisms for managing and cooperating between each MRs are needed.Also the common problem related to all the configuration is where the number of MNP are larger than 1 and at the same time the number of MR or the number of HA or both is larger than one. So a mechanism for solving the *ingress filtering* problem should be used. In most cases the solution is to use second binding on the ingress interface by sending a Prefix-BU through the other MRs and then the HA(s) get(s) all other CoAs.

How do we then distinguish between CoAs.? We use *Preference Settings* One solution is to use an extra identifier for different CoAs and include the identifier information in the update message. This kind of situation exists a lot, except for the cases in which one MNP is only allowed to be controlled by one CoA.

2.6 Policy Based Routing Protocol

Policy is changing the behavior and strategy of a system, without modifying its implementation or interrupting its operation. Policy-based management is largely supported by Standards organizations such as the Internet Engineering Task Force *(IETF)* and the Distributed Management Task Force *(DMTF)* and most network equipment vendors. The focal point in the area of policy-based management is the notion of policy as a means of driving management procedures. Although the technologies for building management building management systems are available, work on the specification and deployment of policies is still scarce.Routing decisions and interface selection are based entirely on IP/network layer information.In order to provide adequate information the level of hierarchy must be considered.The specific information during optimization and deployment is based on the following approach:

Link Layer Information: Interface selection algorithm should take into account all available information and at the same time minimize resource consumption and make decisions with as light computation as possible .However , link quality must be constantlymonitored and the information must be made available for the network layer and user applications in a form that suits them best.

IP layer Information: Several attributes can be retrieved from the IPv6 header without looking into the data, e.g., source address and destination address etc.Some attributes can also be retrieved from IPv6 extension headers (e.g. HOA) only transport protocols like TCP and UDP can be identified directly from the IP header.

Network Originated Information: A service provider may disseminate information about cost, bandwidth and availability of the Internet access in an area using WiMax,WLAN, GPRS and Bluetooth. To advertise such information the default gateway or the access router can send information on cost and bandwidth. The mobile users could then have preferences for connections, like maximize bandwidth or minimize price and the host would select the appropriate interface satisfying these preferences.

This section of the chapter considered the below Algorithm for message notification of mobile node.

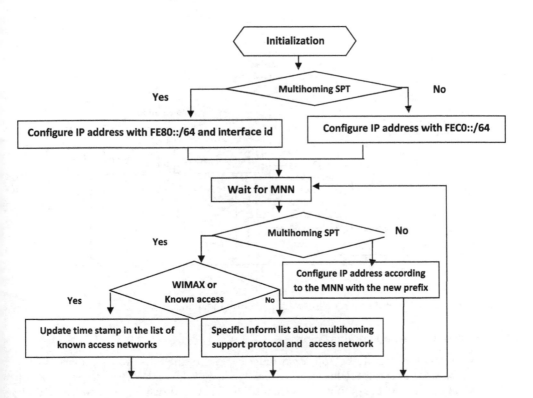

Fig. 4. Algorithm for mobility notification messages at a mobile node during optimization.

2.7 Mechanism for interface selection

The separation of policy and mechanism makes it possible to implement a dynamic interface selection system. The mechanism evaluates connection association and transport information against the actions in policies, using principles.The interface selection system is based on four basic components, *entities, action, policy and mechanism. Entities* define actions. *An entity* may be a user, peer node or 3rd party, e.g., operator. Action is an operation that is defined by an entity and is controlled by the system. *Actions* specify the interfaces to be used for connections on account of entity's requirements. Actions can be presented as conditional statements. Policy governs the actions of an entity. Only one action can take place at a time in a policy. *A policy* set contains several policies possible defined by different entities. *Mechanism* evaluates actions against connection related information and decides which interface is to be used with a specific connection.

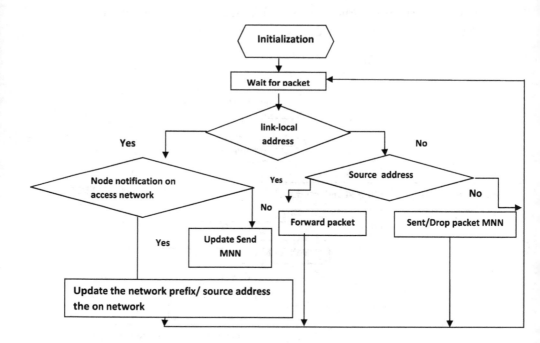

Fig. 5. Algorithm to optimizing the interface selection mechanism.

2.7.1 Network model and optimization

Network model topology to be optimized must contain futures that addressed parameter such as assurance of service delivery and security. Since Wimax is a Flexible Access Point System that delivers on the promise of personal broadband and rich service delivery. Paired with a converged IP core and communicating with feature-rich, multimodal devices combining one network, one service delivery platform and seamless experience that is transparent to the end users. The below fig 6 consider the network model topology for optimization of IPv6 over wimax.

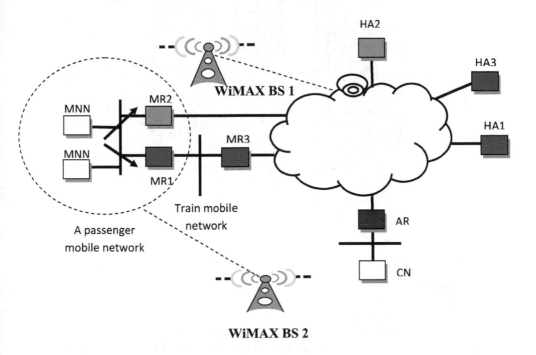

Fig. 6. Network Model Topology For Optimization of IPv6 over WIMAX

From the Network model the wimax BS1, wimax BS2,AR was coined from the architectural specification that depict the concept of Wimax deployment. The access Service Network (ASN) mainly was used for regrouping of BS and AR.The connectivity service Network (CSN) offers connectivity to the internet. To optimize using policy based routing protocol. The link layer information,IP layer information,Network originated information are initialized.

2.8 Standard for WiMAX architecture

WiMAX is a term coined to describe standard, interoperable implementations of IEEE 802.16 wireless networks, similar to the way the term Wi-Fi is used for interoperable implementations of the IEEE 802.11 Wireless LAN standard. However, WiMAX is very different from Wi-Fi in the way it works. The architecture defines how a WiMAX network connects with other networks, and a variety of other aspects of operating such a network, including address allocation, authentication. An overview of this specification for different architectures in order to deploy IPv6 over WiMAX is depicted below in Fig 7 by WiMAX forum .

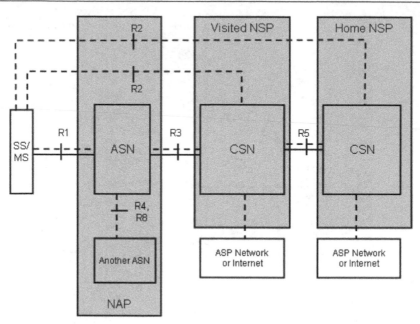

Fig. 7. Architectural Specification for Deployment of IPv6 over WiMAX.

In the proposed network model and optimization the reminder should note that : Regrouping of BS and AR into one entity is named the Access Service Network (ASN) for WiMAX. It has a complete set of functions such as AAA (Authentication, Authorization, Accounting), Mobile IP Foreign agent, Paging controller, and Location Register to provide radio access to a WiMAX Subscriber. The Connectivity Service Network (CSN) offers connectivity to the internet. In the ASN, the BS and AR (or ASN-Gateway) are connected by using either a Switch or Router. The ASN has to support Bridging between all its R1 interfaces and the interfaces towards the network side; forward all packets received from any R1 to a network side port and flood any packet received from a network side port destined for a MAC broadcast or multicast address to all its R1 interfaces. The SS are now considered as mobile (MS), the support for dormant mode is now critical and a necessary feature. Paging capability and optimizations are possible for paging an MS are neither enhanced nor handicapped by the link model itself. However, the multicast capability within a link may cause for an MS to wake up for an unwanted packet.

The solution can consist of filtering the multicast packets and delivering the packets to MS that are listening for particular multicast packets. To deploy IPv6 over IEEE 802.16, SS enters the networks and auto-configure its IPv6 address. In IEEE 802.16, when a SS enters the networks it gets three connection identifier (CID) connections to set-up its global configuration. The first CID is usually used for transferring short, sensitive MAC and radio link control messages, like those relating to the choice of the physical modulations. The second CID is more tolerant connection, it is considered as the primary management connection. With this connection, authentication and connection set-up messages are exchanged between SS and BS. Finally, the third CID is dedicated to the secondary management connection.

2.8.1 WiMAX security

WiMAX Security is a broad and complex subject most especially in wireless communication networks. The subject mechanism of Wimax Technology must meet the requirement design for security architecture in Wimax. Each layer handles different aspects of security, though in some cases, there may be redundant mechanisms. As a general principle of security, it is considered good to have more than one mechanism providing protection so that security is not compromised in case one of the mechanisms is broken. Security goals for wireless networks can be summarized as follows. Privacy or confidentiality is fundamental for secure communication, which provides resistance to interception and eavesdropping.

Message authentication provides integrity of the message and sender authentication, corresponding to the security attacks of message modification and impersonation. Anti-replay detects and disregards any message that is a replay of a previous message. Non-repudiation is against denial and fabrication. Access control prevents unauthorized access. Availability ensures that the resources or communications are not prevented from access by DoS attack. The 802.16 standard specifies a security sub layer at the bottom of the MAC layer. This security sub layer provides SS with privacy and protects BS from service hijacking. There are two component protocols in the security sub layer: an encapsulation protocol for encrypting packet data across the fixed BWA, and a Privacy and Key Management Protocol (PKM) providing the secure distribution of keying data from BS to SS as well as enabling BS to enforce conditional access to network services. The model below was adapted based on security in wimax. This chapter is still investigating the protocol in the sub layer that can mitigate encapsulation of packet data.

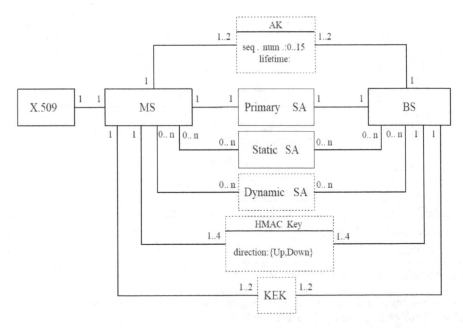

Fig. 8. Security model

2.8.2 Authentication in WiMAX

Basically in WIMAX/802.16 there are three options for authentication: device list based, X.509 based or EAP-based. If device list-based authentication is used only, then the likelihood of a BS or MA masquerading attack is likely. Impact can be high. The risk is therefore high and there is a need for countermeasures. If X.509-based authentication is used, the likelihood for a user (a MS) to be the victim of BS masquerading is possible because of the asymmetry of the mechanism.

There are specific techniques that identify theft and BS attack. Identity theft consists of reprogramming a device with the hardware address of another device. This is a well know problem in unlicensed services such as WiFi/802.11, but in cellular networks because it had been made illegal and more difficult to execute with subscriber ID module (SIM) cards. The exact method of attack depends on the type of networks.

The proposed policy based routing protocol for optimization evaluates connection association and transport information against the actions in policies, using the following principles:

- The mechanism must allow dynamic management of policies such as add, update and remove operations.
- The evaluation of policies should always result in exactly one interface for any traffic flow or connection. This is reached by having a priority order for actions.
- All attribute-value pairs in an action must match for a traffic flow or connection for the action to take place.
- The mechanism selects an interface based on the priority order of interfaces in an action.
- The mechanism uses default actions which match to all flows and connections if no other matching action is found. The mechanism should support distributed policy management and allow explicit definition of priorities. The below table consider our model for optimization during Authentication of WIMAX/802.16.

Threat	DoS on BS or MS
Kind	Mechanism
Device	Device List : RSA / X.509 Certificate
User Level	EAP + EAP – TLS (X.509) or EAP – SIM (subscriber ID module)
Data Traffic	AES-CCM CBC-MAC
Physical Layer Header	None
MAC Layer Header	None
Management messages	SHA – 1 Based MAC AES Based MAC

Table 2. Authentication in WIMAX 802.16

The intended proposed concept is to mitigate and prevent Dos on the BS or MS by introducing **Policy Repository.** In a WiMax/802.16 network, it is more difficult to do these because of the time division multiple access model. The attacker must transmit while the legitimate BS is transmitting. The signal of the attacker, however, must arrive at the targeted receiver MS(s) with more strength and must put the signal of the legitimate BS in the background, relatively speaking. Again, the attacker has to capture the identity of a legitimate BS and to build a message using that identity. The attacker has to wait until a time slot allocated to the legitimate BS starts. The attacker must transmit while achieving a receive signal strength. The receiver MSs reduce their gain and decode the signal of the attacker instead of the one from the legitimate BS.

2.8.3 Key management in Wimax for 5G technology

5G technology has a bright future because it can handle best technologies. The primary concern that should be focused on in 5G is the automated and optimization capability to support software. Although the issue of handover is being address since the Router and switches in this Network has high connectivity capability. Security is under studied in this regard. The knowledge base for the key management for 5G technology centered on physical layer, privacy sub layer threat, mutual Authentication, Threat of identity theft, water Torture and Black hat threat in wimax technology. The protocol used is not rolled out because some flexible framework created by the IETF (RFC 3748), allows arbitrary and complicated **authentication protocols** to be exchanged between the supplicant and the authentication server. **Extensible Authentication Protocol** (EAP) is a simple encapsulation that can run over not only PPP but also any link, including the WiMAX link. A number of **Extensible Authentication Protocol** (EAP) methods have already been defined to support authentication, using a variety of credentials, such as passwords, certificates, tokens, and smart cards. For example, **Protected Extensible Authentication Protocol** (PEAP) defines a password- based EAP method, EAP-transport-layer security (EAP-TLS) defines a certificate-based **Extensible Authentication Protocol** (EAP) method, and EAP-SIM (subscriber identity module) defines a SIM card–based EAP method. EAP-TLS provides strong mutual authentication, since it relies on certificates on both the network and the subscriber terminal.(Chong li).

3. Conclusion

Considering the complex issues and areas that have been addressed in this book chapter. The main focus of the chapter is how to provide techniques on automation and optimization using Algorithm based on policy based routing protocol. However, the various issues on this subject matter have been addressed. Analysis of micro mobility via Macro mobility based on comparative investigation and requirement was advanced. Furthermore the key optimization and data communication of IPv6 over wimax deployment must consider: *session preservation and interface selection mechanism.* The account of policy based routing protocol must provide: *link layer information, IP layer information and network originated information.* Our network model topology for optimization evaluates connection association and transport information against the actions in the policies using the aforementioned Algorithm. The remainder of this report should note that there are limitations in wimax deployment such as: low bit rate, speed of connectivity and sharing of bandwidth.

Finally, the chapter provides the basic Algorithm for optimization of IPv6 over wimax deployment using policy based routing protocol.

4. References

[1] Mobile IPv6 Fast Handovers over IEEE 802.16e Networks H. Jang, J. Jee, Y. Han, S. Park, J. Cha, June 2008

[2] IPv6 Deployment Scenarios in 802.16 Networks M-K. Shin, Ed., Y-H. Han, S-E. Kim, D. Premec, May 2008.

[3] Transmission of IPv6 via the IPv6 Convergence Sublayer over IEEE 802.16 Networks B. Patil, F. Xia, B. Sarikaya, JH. Choi, S. Madanapalli, February 2008.

[4] Mobility Support in IPv6 D. Johnson, C. Perkins, J. Arkko,RFC 3775. June 2004.

[5] Analysis of IPv6 Link Models for 802.16 Based Networks S. Madanapalli, Ed. ,August 2007.

[6] Threats Relating to IPv6 Multihoming Solutions E. Nordmark, T. Li, October 2005.

[7] Koodli, "Fast Handovers for Mobile IPv6", IETF RFC- 4068, July 2005.

2

Towards Semantic Interoperability in Information Technology: On the Advances in Automation

Gleison Baiôco, Anilton Salles Garcia and Giancarlo Guizzardi

Federal University of Espírito Santo

Brazil

1. Introduction

Automation has, over the years, assumed a key role in various segments of business in particular and, consequently, society in general. Derived from the use of technology, automation can reduce the effort spent on manual work and the realization of activities that are beyond human capabilities, such as speed, strength and precision. From traditional computing systems to modern advances in information technology (IT), automation has evolved significantly. At every moment a new technology creates different perspectives, enabling organizations to offer innovative, low cost or custom-made services. For example, advents such as artificial intelligence have enabled the design of intelligent systems capable of performing not only predetermined activities, but also ones involving knowledge acquisition. On the other hand, customer demand has also evolved, requiring higher quality, lower cost or ease of use. In this scenario, advances in automation can provide innovative automated services as well as supporting market competition in an effective and efficient way. Considering the growing dependence of automation on information technology, it is observed that advances in automation require advances in IT.

As an attempt to allow that IT delivers value to business and operates aligned with the achievement of organizational goals, IT management has evolved to include IT service management and governance, as can be observed by the widespread adoption of innovative best practices libraries such as ITIL (ITIL, 2007) and standards such as ISO/IEC 20000 (ISO/IEC, 2005). Nonetheless, as pointed out by Pavlou & Pras (2008), the challenges arising from the efforts of integration between business and IT remain topic of various studies. IT management, discipline responsible for establishing the methods and practices in order to support the IT operation, encompasses a set of interrelated processes to achieve this goal. Among them, configuration management plays a key role by providing accurate IT information to all those involved in management. As a consequence, semantic interoperability in the domain of configuration management has been considered to be one of the main research challenges in IT service and network management (Pras et al., 2007). Besides this, Moura et al. (2007) highlight the contributions that computer systems can play in terms of process automation, especially when they come to providing intelligent solutions, fomenting self-management. However, as they emphasize, as an emerging paradigm, this initiative is still a research challenge.

According to Pras et al. (2007), the use of ontologies has been indicated as state of the art for addressing semantic interoperability, since they express the meaning of domain concepts and relations in a clear and explicit way. Moreover, they can be implemented, thereby enabling process automation. In particular, ontologies allow the development of intelligent systems (Guizzardi, 2005). As a result, they foment such initiatives as self-management. Besides this, it is important to note that ontologies can promote the alignment between business and IT, since they maximize the comprehension regarding the domain conceptualization for humans and computer systems. However, although there are many works advocating their use, there is not one on IT service configuration management that can be considered as a de facto standard by the international community (Pras et al., 2007).

As discussed in Falbo (1998), the development of ontologies is a complex activity and, as a result, to build high quality ontologies it is necessary to adopt an engineering approach which implies the use of appropriate methods and tools. According to Guizzardi (2005, 2007), ontology engineering should include phases of conceptual modeling, design and implementation. In a conceptual modeling phase, an ontology should strive for expressiveness, clarity and truthfulness in representing the domain conceptualization. These characteristics are fundamental quality attributes of a conceptual model responsible for its effectiveness as a reference framework for semantic interoperability. The same conceptual model can give rise to different ontology implementations in different languages, such as OWL and RDF, in order to satisfy different computational requirements. Thus, each phase shall produce different artifacts with different objectives and, as a consequence, requires the use of languages which are appropriate to the development of artifacts that adequately meet their goals. As demonstrated by Guizzardi (2006), languages like OWL and RDF are focused on computer-oriented concerns and, for this reason, improper for the conceptual modeling phase. Philosophically well-founded languages are, conversely, committed to expressivity, conceptual clarity as well as domain appropriateness and so suitable for this phase.

Considering these factors, Baiôco et al. (2009) present a conceptual model of the IT service configuration management domain based on foundational ontology. Subsequently, Baiôco & Garcia (2010) present an implementation of this ontology, describing how a conceptual model can give rise to various implementation models in order to satisfy different computational requirements. The objective of this chapter is to provide further details about this IT service configuration management ontology, describing the main ontological distinctions provided by the use of a foundational ontology and how these distinctions are important to the design of models aligned with the universe of discourse, maximizing the expressiveness, clarity and truthfulness of the model and consequently the semantic interoperability between the involved entities. Moreover, this chapter demonstrates how to apply the entire adopted approach, including how to generate different implementations when compared with previous ones. This attests the employed approach, makes it more tangible and enables to validate the developed models as well as demonstrating their contributions in terms of activity automation.

It is important to note that the approach used in this work is not limited to the domain of IT service configuration management. In contrast, it has been successfully employed in many fields, such as oil and gas (Guizzardi et al., 2009) as well as medicine (Gonçalves et al., 2011). In fact, the development of a computer system involves the use of languages able to adequately represent the universe of discourse. According to Guizzardi (2005), an imprecise representation of state of affairs can lead to a false impression of interoperability, i.e.

although two or more systems seem to have a shared view of reality, the portions of reality that each of them aims to represent are not compatible. As an alternative, ontologies have been suggested as the best way to address semantic interoperability. Therefore, in particular, the ontological evaluation realized in this work contributes to the IT service configuration management domain, subsidizing solutions in order to address key research challenges in IT management. In general, this chapter contributes to promote the benefits of the employed approach towards semantic interoperability in IT in various areas of interest, maximizing the advances in automation. Such a contribution is motivated in considering that although recent research initiatives such as that of Guizzardi (2006) have elaborated on why domain ontologies must be represented with the support of a foundational theory and, even though there are many initiatives in which this approach has been successfully applied, it has not yet been broadly adopted. As reported by Jones et al. (1998), most existing methodologies do not emphasize this aspect or simply ignore it completely, mainly because it is a novel approach.

In this sense, this chapter is structured as follows: Section 2 briefly introduces the IT service configuration management domain. Section 3 discusses the approach to ontology development used in this work. Section 4 presents the conceptual model of the IT service configuration management domain. Section 5 shows an implementation model of the conceptual model presented in Section 4 and finally Section 6 relates some conclusions and future works.

2. IT service configuration management

The business of an organization requires quality IT services economically provided. According to ITIL, to be efficient and effective, organizations need to manage their IT infrastructure and services. Configuration management provides a logical model of an infrastructure or service by identifying, controlling, maintaining and verifying the versions of configuration items in existence. The logical model of IT service configuration management is a single common representation used by all parts of IT service management and also by other parties, such as human resources, finance, suppliers and customers. A configuration item, in turn, is an infrastructure component or an item that is or will be under the control of configuration management (ITIL, 2007; ISO/IEC, 2005). For innovative IT management approaches such as ITIL and ISO/IEC 20000, configuration items are viewed not only as individual resources but as a chain of related and interconnected resources compounding services. Thus, just as important as controlling each item is managing how they relate to each other. These relationships form the basis for activities such as impact assessment.

According to ITIL and ISO/IEC 20000, a configuration item and the related configuration information may contain different levels of detail. Examples include an overview of all services or a detailed view of each component of a service. Thus, a configuration item may differ in complexity, size and type, ranging from a service, including all hardware, software and associated documentation, to a single software module or hardware component. Configuration items may be grouped and managed together, e.g. a set of components may be grouped into a release. Furthermore, configuration items should be selected using established selection criteria, grouped, classified and identified in such a way that they are manageable throughout the service lifecycle.

As with any process, IT service configuration management is associated with goals that in its case include: (i) supporting, effectively and efficiently, all other IT service management processes by providing configuration information in a clear, precise and unambiguous way; (ii) supporting the business goals and control requirements; (iii) optimizing IT infrastructure settings, capabilities and resources; (iv) subsidizing the dynamism imposed on IT by promoting rapid responses to necessary changes and by minimizing the impact of changes in the operational environment. To achieve these objectives, configuration management should, in summary, define and control the IT components and maintain the configuration information accurately. Based on best practices libraries such as ITIL and standards such as ISO/IEC 20000 for IT service management, the activities of an IT service configuration management process may be summarized as: (i) planning, in order to plan and define the purpose, scope, objectives, policies and procedures as well as the organizational and technical context for configuration management; (ii) identification, aiming to select and identify the configuration structures for all the items (including their owner, interrelationships and configuration documentation), allocate identifiers and version numbers for them and finally label each item and enter it on the configuration management database (CMDB); (iii) control, in order to ensure that only authorized and identified items are accepted and recorded, from receipt to disposal, ensuring that no item is added, modified, replaced or removed without appropriate controlling documentation; (iv) status accounting and reporting, which reports all current and historical data concerned with each item throughout its life cycle; (v) verification and audit, which comprises a series of reviews and audits that verify the physical existence of items and check that they are correctly recorded in the CMDB.

As an attempt to promote efficiency and effectiveness, IT management has evolved to include IT service management and governance, which aims to ensure that IT delivers value to business and is aligned with the achievement of organizational goals. As emphasized by Sallé (2004), in this context, IT processes are fully integrated into business processes. Thus, one of the main aspects to be considered is the impact of IT on business processes and vice versa (Moura et al., 2008). As a consequence, IT management processes should be able to manage the entire chain, i.e. from IT to business. For this reason, the search for the effectiveness of such paradigms towards business-driven IT management has been the topic of several studies in network and service management (Pavlou & Pras, 2008). According to Moura et al. (2007), one of the main challenges is to achieve the integration between these two domains. Configuration management, in this case, should be able to respond in a clear, precise and unambiguous manner to the following question: what are the business processes and how are they related to IT services and components (ITIL, 2007)? Furthermore, as cited by ITIL, due to the scope and complexity of configuration management, keeping its information is a strenuous activity. In this sense, research initiatives consider automation to be a good potential alternative. In fact, the automation of management processes has been recognized as one of the success factors to achieve a business-driven IT management, especially when considering intelligent solutions promoting self-management (Moura et al., 2007). Besides its scope and complexity, configuration management is also closely related to all other management processes. In IT service management and governance, this close relationship includes the interaction among the main entities involved in this context, such as: (i) business, (ii) people, (iii) processes, (iv) tools and (v) technologies (ITIL, 2007). Thus, semantic interoperability among such entities has been characterized as one of the main research challenges, not only in terms of the configuration management process, but in the whole chain of processes that comprise the discipline of network and service management (Pras et al., 2007).

According to Pras et al. (2007), the use of semantic models, in particular, the use of ontologies, has been regarded as the best way with respect to initiatives for addressing issues related to semantic interoperability problems in network and service management. According to these authors, ontologies make the meaning of the domain concepts such as IT management, as well as the relationships between them, explicit. Additionally, this meaning can be defined in a machine-readable format, making the knowledge shared between humans and computer systems, enabling process automation, as outlined by these authors. From this point of view, it is worth mentioning that ontologies are considered as potential tools for the construction of knowledge in intelligent systems (Guizzardi, 2005). Thus, they allow the design of intelligent and above all interoperable solutions, fomenting initiatives as self-management. Finally, it is important to note that ontologies can promote the alignment between business and IT when applied in the context of IT service management and governance since they maximize the expressiveness, clarity and truthfulness of the domain conceptualization for humans and computer systems. However, Pras et al. (2007) point out that despite the efforts of research initiatives, there are still many gaps to be addressed.

Several studies claim that the use of ontologies is a promising means of achieving interoperability among different management domains. However, an ontology-based model and formalization of IT service configuration management remains a research challenge. Regarding limitations, it should be mentioned that ontologies are still under development in the management domain. In fact, the technology is not yet mature and there is not an ontology that can be considered as a *de facto* standard by the international community (Pras et al., 2007). In general, the research initiatives have not employed a systematic approach in the development of ontologies. According to Falbo (1998), the absence of a systematic approach, with a lack of attention to appropriate methods, techniques and tools, makes the development of ontologies more of an art rather than an engineering activity. According to Guizzardi (2005, 2007), to meet the different uses and purposes intended for the ontologies, ontology engineering should include phases of conceptual modeling, design and implementation. Each phase should have its specific objectives and thus would require the use of appropriate languages in order to achieve these goals. However, in most cases, such research initiatives are engaged with the use of technologies and tools such as Protégé and OWL. Sometimes these technologies and tools are used in the conceptual modeling phase, which can result in various problems relating to semantic interoperability, as shown in Guizzardi (2006). At other times, however, they are employed in the implementation phase, ignoring previous phases such as conceptual modeling and design. As a result, such initiatives are obliged to rely on models of low expressivity. Moreover, in most cases, such initiatives propose the use of these technologies and tools for the formalization of network management data models, such as MIB, PIB and the CIM schema. It is noteworthy that data models are closely related to the underlying protocols used to transport the management information and the particular implementation in use. In contrast, information models work at a conceptual level and they are intended to be independent of any particular implementation or management protocol. Working at a higher level, information models usually provide more expressiveness (Pras et al., 2007). Following this approach, Lopez de Vergara et al. (2004) propose an integration of the concepts that currently belong to different network management data models (e.g. MIB, PIB and the CIM schema) in a single model, formalized by ontology languages such as OWL. In an even more specific scenario, i.e. with no intention to unify the various models but only to formalize a particular model, Majewski et al. (2007) suggest the formalization of the CIM schema through ontology languages such as OWL. Similarly, while differentiating the type of data model, Santos (2007) presents an ontology-based network configuration management system. In his work, the proposed ontology was developed according to the MIB data model concepts. As MIB is limited in describing a single system, a view of the entire infrastructure, including the relationships between its components, is not supported by the model. In practice, this gap is often filled by functionalities provided by SNMP-based network management tools which, for example, support the visualization of network topologies (Brenner et al., 2006). Aside from the fact that, in general, the research initiatives are committed to the use of technologies and tools, it is also observed that they are characterized by specific purposes in relation to peculiar applications in information systems that restrict their conceptualizations. In Xu and Xiao (2006), an ontology-based configuration management model for IP network devices is presented, aiming at the use of ontology for the automation of this process. In Calvi (2007), the author presents a modeling of the IT service configuration management described by the ITIL library based on a foundational ontology. The concepts presented and modeled in his work cover a specific need regarding the demonstration of the use of ITIL processes for a context-aware service platform. Finally, there are approaches that seek to establish semantic interoperability among existing ontologies by means of ontological mapping techniques, as evidenced in Wong et al. (2005). However, it is not within the scope of such approaches to develop an ontology but rather to integrate existing ones.

Therefore, in considering the main challenges as well as the solutions which are considered to be state of the art and in analyzing the surveyed works, it is observed that there are gaps to be filled, as highlighted by Pras et al. (2007). In summary, factors such as the adoption of *ad hoc* approaches, the use of inappropriate references about the domain, the intention of specific purposes and, naturally, the integration of existing ontologies, all result in gaps. As a consequence, such factors do not promote the conception of an ontology able to serve as a reference framework for semantic interoperability concerning the configuration management domain in the context of IT service management and governance. This scenario demonstrates the necessity of a modeling which considers the gaps and, therefore, promotes solutions in line with those suggestions regarded as state of the art for the research challenges discussed earlier in this chapter. In particular, it demonstrates the necessity of an appropriate approach for the construction of ontologies as a subsidy for such modeling. In this sense, the next section of this chapter presents an approach for ontology development.

3. Ontology engineering

In philosophy, ontology is a mature discipline that has been systematically developed at least since Aristotle. As a function of the important role played by them as a conceptual tool, their application to computing has become increasingly well-known (Guizzardi et al., 2008). According to Smith and Welty (2001), historically there are three main areas responsible for creating the demand for the use of ontologies in computer science, namely: (i) database and information systems; (ii) software engineering (in particular, domain engineering); (iii) artificial intelligence. Additionally, Guizzardi (2005) includes the semantic web, due to the important role played by this area in the current popularization of the term.

According Guizzardi et al. (2008), an important point to be emphasized is the difference between the senses of the term ontology when used in computer science. In conceptual modeling, the term has been used as its definition in philosophy, i.e. as a philosophically well-founded domain-independent system of formal categories that can be used to articulate domain-specific models of reality. On the other hand, in most other areas of computer science, such as artificial intelligence and semantic web, the term ontology is generally used as: (i) an engineering artifact designed for a specific purpose without giving much importance to foundational issues; (ii) a representation of a particular domain (e.g. law, medicine) expressed in some language for knowledge representation (e.g. RDF, OWL).

From this point of view, the development of ontologies should consider the various uses and, consequently, the different purposes attributed to ontologies as well as any existing interrelationship in order to enable the construction of models that satisfactorily meet their respective goals. However, despite the growing use of ontologies and their importance in computing, the employed development approaches have generally not considered these factors, resulting in inadequate models for the intended purpose. In considering such distinctions Guizzardi (2005, 2007) elaborates and discusses a number of questions in order to elucidate such divergences and thus provide a structured way with respect to the use of ontologies. In addition, Guizzardi and Halpin (2008) describe that the interest in proposals for foundations in the construction of ontologies has been the topic of several studies and they report some innovative and high quality research contributions. It is based on such questions that are elaborated the further discussions contained in this section and thus the approach used for the construction of the ontological models proposed in this work.

As discussed in Falbo (1998), the development of ontologies is a complex activity and, hence, in order to build high quality ontologies, able to adequately meet their various uses and purposes, it is necessary to adopt an engineering approach. Thus, unlike the various *ad hoc* approaches, the construction of ontologies must use appropriate methods and tools. Falbo (2004) proposes a method for building ontologies called SABiO (Systematic Approach for Building Ontologies). This method proposes an life cycle by prescribing an iterative process that comprises the following activities: (i) purpose identification and requirements specification, which aims to clearly identify the ontology's purpose and its intended use by means of competence questions; (ii) ontology capture, viewing to capture relevant concepts existing within the universe of discourse as well as their relationships, properties and constraints, based on the competence questions; (iii) ontology formalization, which is responsible for explicitly representing the captured conceptualization by means of a formal language, such as the definition of formal axioms using first-order logic; (iv) integration with existing ontologies, in order to search for other ones with the purpose of reuse and integration; (v) ontology evaluation, which aims to identify inconsistencies as well as verifying truthfulness in line with the ontology's purpose and requirements; (vi) ontology documentation. Noticeably, the competence questions form an important concept within SABiO, i.e. the questions the ontology should be able to answer. They provide a mechanism for defining the scope and purpose of the ontology, guiding its capture, formalization and evaluation - regarding this last aspect, especially with respect to the completeness of the ontology.

The elements that constitute the relevant concepts of a given domain, understood as domain conceptualization, are used to articulate abstractions of certain states of affairs in reality, denominated as domain abstraction. As an example, consider the domain of product sales.

A conceptualization of this domain can be constructed by considering concepts such as, *inter alia*: (i) customer, (ii) provider, (iii) product, (iv) is produced by, (v) is sold to. By means of these concepts, it is possible to articulate domain abstractions of certain facts extant in reality, such as: (i) a product is produced by the provider and sold to the customer. It is important to highlight that conceptualizations and abstractions are abstract entities which only exist within the mind of a user or a community of users of a language. Therefore, in order to be documented, communicated and analyzed, they must be captured, i.e. represented in terms of some concrete artifact. This implies that a language is necessary for representing them in a concise, complete and unambiguous way (Guizzardi, 2005). Figure 1-a presents "Ullmann's triangle" (Ullmann, 1972), which illustrates the relation between a language, a conceptualization and the part of reality that this conceptualization abstracts. The relation "represents" concerns the definition of language semantics. In other words, this relation implies that the concepts are represented by the symbols of language. The relation "abstracts", in turn, denotes the abstraction of certain states of affairs within the reality that a given conceptualization articulates. The dotted line between language and reality highlights the fact that the relation between language and reality is always intermediated by a certain conceptualization. This relation is elaborated in Figure 1-b, which depicts the distinction between an abstraction and its representation, as well as their relationships with the conceptualization and representation language. The representation of a domain abstraction in terms of a representation language is called model specification (or simply model, specification or representation) and the language used for its creation is called modeling language (or specification language).

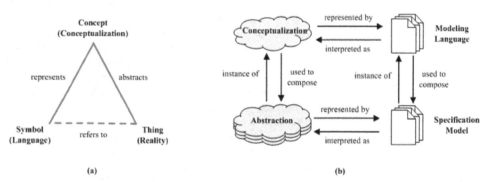

(a)   (b)

Fig. 1. Ullmann's triangle and relations between conceptualization, abstraction, modelling language and model, according to Guizzardi (2005).

Thus, in addition to the adoption of appropriate methods, able to systematically lead the development process, ontology engineering as an engineering process aims at the use of tools, which should be employed in accordance with the purpose of the product that is being designed. In terms of an ontology development process such tools include modeling languages or even ontology representation languages. According to Guizzardi (2005), one of the main success factors regarding the use of a modeling language is its ability to provide its users with a set of modeling primitives that can directly express the domain conceptualization. According to the author, a modeling language is used to represent a conceptualization by compounding a model that represents an abstraction, which is an instance of this conceptualization. Therefore, in order for the model to faithfully represent an abstraction, the modeling primitives of the language used to produce the model must accurately represent the domain conceptualization used to articulate the abstraction represented by the model. According to Guizzardi (2005), if a conceptual modeling language is imprecise and coarse in the description of a given domain, then there can be specifications of the language which, although grammatically valid, do not represent admissible state of affairs. Figure 2-a illustrates this situation. The author also points out that a precise representation of a given conceptualization becomes even more critical when it is necessary to integrate different independently developed models (or systems based on these models). As an example, he mentions a situation in which it is necessary to have the interaction between two independently developed systems which commit to two different conceptualizations. Accordingly, in order for these two systems to function properly together, it is necessary to ensure that they ascribe compatible meanings to the real world entities of their shared subject domain. In particular, it is desirable to reinforce that they have compatible sets of admissible situations whose union (in the ideal case) equals the admissible states of affairs delimited by the conceptualization of their shared subject domain. The ability of entities (in this case, systems) to interoperate (operate together) while having compatible real-world semantics is known as semantic interoperability (Vermeer, 1997). Figure 2-b illustrates this scenario.

Fig. 2. Consequences of an imprecise and coarse modelling language (Guizzardi, 2005).

In Figure 2-b, C_A and C_B represent the conceptualizations of the domains of systems A and B, respectively. As illustrated in this figure, these conceptualizations are not compatible. However, because these systems are based on poor representations of these conceptualizations, their sets of considered possible situations overlap. As a result, systems A and B agree exactly on situations that are neither admitted by C_A nor by C_B. In summary, although these systems appear to have a shared view of reality, the portions of reality that each of them aims to represent are not compatible. Therefore, the more it is known about a given domain and the more precisely it is represented, the bigger the chance of obtaining interpretations that are consistent with the reality of that domain and, therefore, of achieving semantic interoperability between the entities involved in these interpretations. Thus, Guizzardi (2005) concludes that, on the one hand, a modeling language should be sufficiently expressive to adequately characterize the conceptualization of the domain and, on the other hand, the semantics of the produced specifications should be clear, allowing users to recognize what language constructs mean in terms of domain concepts. Moreover, the specification produced by means of the language should facilitate the user in understanding and reasoning about the represented state of affairs.

In view of the different purposes, Guizzardi (2005, 2007) highlights that ontology engineering, analogous to software engineering and information systems, must include phases of conceptual modeling, design and implementation. Each phase has its specific objectives and thus requires different types of methods and tools to meet its particular characteristics. As mentioned, during a conceptual modeling phase, an ontology must strive for expressivity, clarity and truthfulness in representing the domain conceptualization. Therefore, the conceptual modeling phase requires specialized languages so as to create ontologies that approximate as closely as possible to the ideal representation of the domain. The same conceptual model can give rise to different implementation models in different languages, such as OWL and RDF, in order to satisfy different non-functional requirements, such as decidability and completeness. The section delimited as *Level* in Figure 3 illustrates this approach based on relations between conceptualization, abstraction, modeling language and model, shown in Figure 1-b. According to Guizzardi (2006), semantic web languages such as OWL and RDF are focused on computation-oriented concerns and are therefore inadequate for the conceptual modeling phase. Philosophically well-founded languages, on the other hand, are engaged in expressivity, conceptual clarity and domain appropriateness and are therefore suitable for this phase. To support his assertion, Guizzardi (2006) presents several problems of semantic interoperability from the use of semantic web languages in the representation of the domain and demonstrates how philosophically well-founded languages are able to address these problems.

As shown in Guizzardi (2005), while domain conceptualizations and, consequently, domain ontologies are established by the consensus of a community of users with respect to a material domain, a conceptual modeling language (which can be used to express these domain ontologies) must be rooted in a domain independent system of real-world categories, philosophically and cognitively well-founded, i.e. a foundational ontology. Foundational ontologies aggregate contributions from areas such as descriptive metaphysics, philosophical logic, cognitive science and linguistics. The theories inherent to these areas are called (meta-) conceptualizations and describe knowledge about reality in a way which is independent of language and particular states of affairs. A foundational ontology, in turn, is the representation of these theories in a concrete artifact. Thus, foundational ontologies, in the philosophical sense, can be used to provide real-world semantics for modeling languages as well as to constrain the possible interpretations of their modeling primitives, increasing the clarity of interpretation and, consequently, reducing ambiguities (which are key success factors in achieving semantic interoperability). Accordingly, it is possible to build domain ontologies by means of conceptual modeling languages based on foundational ontologies. In this sense, the *Meta-level* section in Figure 3, in addition to the *Level* section, represents the approach proposed by Guizzardi (2005, 2007).

An example of a foundational ontology is UFO (Unified Foundational Ontology). UFO was initially proposed in Guizzardi and Wagner (2004) and its most recent version is presented by Guizzardi et al. (2008). It is organized in three incrementally layered compliance sets: (i) UFO-A, which is essentially the UFO's core, defining terms related to endurants (objects, their properties etc); (ii) UFO-B, which defines as an increment to UFO-A terms related to perdurants (events etc); (iii) UFO-C, which defines as an increment to UFO-B terms explicitly related to the spheres of social entities.

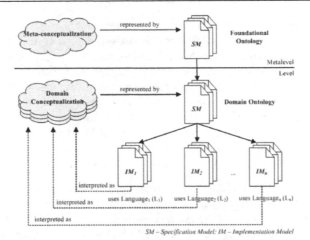

Fig. 3. Ontology engineering approach proposed by Guizzardi (2005, 2007).

As a conclusion, the ontology engineering approach described in this section considers the distinctions of the term ontology as well as their interrelationships, thereby establishing the phases, their respective objectives, as well as the methods and tools appropriate for the characteristic of each phase, allowing thus the construction of models capable of meeting the various purposes intended for them. Therefore, based on the study about IT service configuration management, as well as the study about ontology development, it is possible to construct an ontology of this domain as the objective of this chapter.

4. Conceptual model of the IT service configuration management domain

In considering the main research challenges, as well as the initiatives for solutions which are regarded state of the art, which properly lead the identification of gaps, as much as the appropriated approach to fulfil them, this section presents the conceptual models proposed in this work. On the basis of Figure 3, which shows the ontology development approach adopted in this work, the conceptual model proposed in this section concerns the domain ontology, whose conceptualization in discussion is the IT service configuration management. Regarding the foundational ontology, this work uses UFO, which represents the meta-conceptualization responsible for promoting the philosophical base of the work.

As discussed in Section 2, configuration management is responsible for maintaining information about configuration items and providing them to all the other management processes. In the context of IT service management and governance, configuration management must be able to answer questions such as: what are the business processes and how do they relate to the IT services and components? Based on this question and given that the main goal of this ontology is to describe a theory of the domain of IT service configuration management independent of specific applications, the defined competency questions reflect this intention. In this case, they lead to a mapping between IT and business concepts, as follows:

CQ1: How do the IT services and the business processes of an organization relate?

CQ2: How do the IT services and the IT components such as hardware and software relate?

To answer these questions, it is necessary to address others, such as: (i) what is an IT service? (ii) what is a business process? For this reason, Baiôco et al. (2009) propose an IT service configuration management ontology, which addresses such questions in order to provide a basis for the ontology presented in this chapter. As described through the SABiO method, if the domain of interest is too complex, a decomposing mechanism should be used in order to better distribute this complexity. In this case, a potentially interesting approach is to consider sub-ontologies. Therefore, to answer the competency questions, the following sub-ontologies were developed: (i) business process; (ii) IT service; (iii) IT component and lastly (iv) configuration item. These sub-ontologies complement each other in constituting the IT service configuration management ontology discussed in this work.

In terms of reusing existing ontologies, it is important to mention that besides the adopted literature, the conceptual modeling of this section also takes into consideration the discussions inherent to processes in general done in Falbo (1998) and Guizzardi et al. (2008), as well as the discussions inherent to IT services done in Calvi (2007) and Costa (2008). Still in line with the SABiO method, during the capture of the ontology the use of a graphic representation is essential to facilitate the communication between ontology engineers and domain experts. However, a graphical model is not enough to completely capture an ontology. This way, axioms should be provided to reinforce the semantics of the terms and establish the domain restrictions. Thus, the sub-ontologies developed in this work are connected by relations between their concepts and by formal axioms. Due to limitations of space, will be shown only those axioms also used in the next section. To distinguish the subject domain concepts and the UFO concepts, these last ones are presented in blank in the conceptual model that follows. Figure 4 presents part of the proposed ontology.

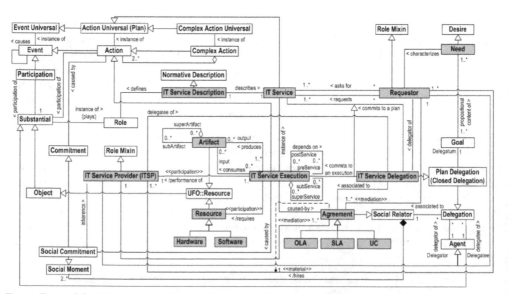

Fig. 4. Part of the IT service configuration management ontology.

According to the ITIL library, an IT service is a service which is provided by an IT organization to one or more clients. Therefore, in accordance with the UFO terminology, an IT service is characterized as a type of plan, i.e. an intentional event. As a result, the properties inherent to events are applied to IT services. As established in UFO, events are possible changes from one portion of reality to another, which means they can transform reality by altering the state of affairs from a pre-state to a post-state. Consequently they can produce, direct or indirectly, situations that satisfy the necessary conditions for other events to happen. On that account, events can cause other events, including then the service executions as represented in the model by the relation causes. An IT service execution (ITSE), in turn, denotes one or more particular actions that occur at specific time intervals, aiming to satisfy the propositional content of a commitment. On this note, an IT service execution is an action that instantiates a type of plan, in this case an IT service. This distinction, derived from the foundational ontology, provides greater adequacy to the domain, making it possible to distinguish services from their executions and allowing the comparison, for example, between achieved and planned results.

Because it denotes one or more actions, an IT service execution can be atomic or complex. As a complex execution, it is decomposed into other smaller service executions, termed subservices. In this way, a subservice is a service execution that is part of a bigger service execution, its super-service. As the properties inherent to events are applied to the IT services, the decomposition of service executions, as any other event decomposition, is characterized as a transitive, asymmetric and irreflexive relation. This inheritance of properties from the foundational ontology facilitates modeling decisions and minimizes the possibilities of incoherent descriptions of the domain.[1]

According to the ITIL library, an IT service execution aims to produce resources in order to satisfy the needs of its customers. On the conceptual model proposed in this section, the produced resources are said to be outputs. On the other hand, an IT service execution can consume resources, seen as raw material, to produce results. On the model, these resources are said to be inputs. From the point of view of UFO, an artifact of a service execution is a type of resource (UFO::Resource) which in turn is mapped to the notion of an object. As such, the subartifact and superartifact relations are then governed by the axioms defined for the (different types of) parthood relations between substantials, as described in Guizzardi (2005).

According to UFO, a resource (UFO::Resource) is a role that an object plays in an event. Thus, the artifacts of a service execution are roles played by objects in the scope of this service instance. This being said, it is important to highlight a contribution from UFO attributed to the model. As the notions of objects and roles are defined, it becomes possible to represent real situations of the domain, as with those where the same object plays the role of an output to a service execution and input to another, distinguishing only the type of participation performed by the object and keeping its identity throughout its existence. This is because, according to UFO, an object is a type of endurant which, in contrast to a

---

[1] UFO makes explicit distinctions often ignored by many languages. For example, while it is possible to consider that an event x is a part of an event z because x is a part of an event y that is a part of z, it is not the case that the musician's hand (and so a part thereof) is a part of the band within which the musician is a part. In the first case, there is transitivity, but in the latter this does not exist. In this sense, parthood relations denote distinctions that should be considered.

perdurant (e.g. an event), is an individual that keeps its identity throughout its existence. On the model, the participation of the output is represented through the relation "produces" while the participation of the input is represented through the relation "consumes".

As objects, the participation of artifacts in a service execution should correspond to the types of participation of objects in an action, defined in UFO. In fact, being produced or consumed does not express the exact notion on the participation of an object in a service execution. Namely, outputs can be created or modified. Inputs, on the other hand, can be used, modified into new products or else terminated. As such, the relation "produces", as well as the relation "consumes", designates distinct notions which should be considered. Therefore, in light of the UFO terminology and the subject domain, the relation "produces" denotes the participation of creation or change, while the relation "consumes" indicates the participation of usage, change or termination as defined in UFO. This demonstrates the foundation of the domain concepts and relations in terms of philosophically well-founded concepts and relations, making the representation of the universe of discourse even more clear, expressive and coherent with reality.

As described by the ITIL library, an IT service is based on the use of the information technology, which includes, among other things, IT components such as hardware and software. Therefore, an IT service execution, as any other activity, presumes the use of resources, in particular IT components, in order to achieve results. Essentially, under the UFO's perspective, the resources of a service execution are types of resources (UFO::Resource), i.e. objects participating in an action. Therefore, not only artifacts but also resources are roles played by objects within the scope of a service execution. Again, the distinction between objects and roles contributes to the representation of real situations in the universe of discourse, including those where the same object is produced by a service execution but is required by another, though remaining as the same individual. In the model, the participation of a resource is represented by the relation "requires". Once these resources are used as support tools in a service execution, the foundation associated with this relation leads to only one of the types of resource participation defined in UFO, i.e. the usage participation.

In considering the definitions of artifacts and resources, especially regarding the foundation provided by UFO, it is noted that the inputs and the resources are objects which can play the same type of participation in a service execution, i.e. the usage participation. However, taking a service execution as a transformation primitive, inputs indicate raw materials which are incorporated into the product. Resources, on the other hand, refer to components that support a service execution, but are not intended as products of this execution. Thus, the resources employed in a service execution cannot be considered as products within the scope of this execution. Therefore, by clearly representing the participation of inputs and resources in a service execution, this foundation promoted the identification with regard to the similar type of participation of these roles. As a consequence, it required the use of domain definitions able to characterize such roles, since fundamentally they are similar. Hence, as pointed out, this observation becomes evident, in this case, especially due to the foundation of the domain concepts in terms of UFO. This demonstrates the contribution of a foundational ontology, in general, and of UFO, in particular, in supporting the construction of appropriate models regarding domain and comprehension.

As described in Costa (2008), an IT service execution is an intended and orderly execution of one or more actions in order to satisfy the propositional content (goal) of a commitment agreed with an agent. This ordering, in some cases, is governed according to certain situations that culminate into the execution of the service. In this sense, the causality relation between events leads to a dependence relation between them. This happens because, by changing the state of affairs of the reality from one state (pre-state) to another (post-state), events can generate (directly or indirectly) a situation which satisfies a necessary condition for other events to occur. In summary, an event depends on another if the first is caused, directly or indirectly, by the second. Hence, service executions can depend on events (or on other service executions) in order to occur. It is relevant to mention that artifacts and resources can imply a dependency relation between service executions. In fact, as described in UFO, situations are complex entities that agglutinate other entities (including objects) and denote the pre- and post-state of an event. So, as a type of object, artifacts and resources are present in situations that symbolize the pre- and post-states of service executions. As an example, a service may require resources or consume inputs that are products of other services, characterizing thus a dependence relation between them because of their respective objects. Besides this, it is important to highlight that if an event depends on the other then there is a temporal relation between them that should be taken into account. Concerning dependence between service executions, this temporal relation is represented by the pre-service and post-service relations. As the causality relation between events, the dependence relation between them (including the IT service executions) is transitive, asymmetric and irreflexive. These properties are also valid for the relations pre- and post-service. Temporal relations between events make possible to define the temporal ordering in which the events (including the service executions) are submitted and, therefore, establish the order in which they occur, even when there is no dependence relation defined between them. This is because, according to UFO, events are framed into temporal intervals, from which originate the temporal relations. Thus, the model proposed in this section is based on a framework of concepts and relations defined in UFO which makes possible to specify the flow in which the events are associated and, consequently, the ordering associated with the IT service executions. In this way, this structure supports the modeling decisions at the same time as it contributes to the creation of a more expressive, clear and truthful model with regard to the universe of discourse.

An IT service is described by a normative description, termed IT service description. The description of an IT service, for instance, describes the roles played by each agent in a service execution. Agents, as well as objects, are substantial from the UFO point of view. However, agents differ from objects because of the fact that they can possess beliefs, desires and intentions. Intentions are characterized as desired states of affairs for which the agent commits itself to pursuing, i.e. an internal commitment. For this reason, intentions cause the agent to perform actions. In this sense, the participation of an agent in an action is characterized as an action contribution. Consequently, service executions are performed by agents. Indeed, the action contribution of an agent in a service execution is caused by a social commitment of the agent in performing this service execution (or part thereof, the subservice) with its consequent permissions and obligations. Therefore, the role modeling pattern described by UFO applies to the IT service domain. In fact, as advanced in Guizzardi (2006), as a domain independent knowledge representation language, foundational ontologies in general and UFO in particular aim to support the construction of domain-dependent models by acting as a reference framework and thus guiding modeling decisions and allowing the creation of models that clearly and accurately represent, as much as possible, the real situations in a universe of discourse. This pattern has resolved various role modeling problems in the literature. Thus, the execution of a service instance is characterized by an agent playing a role in an IT service, in this case, acting as a service provider. It is worth noting that not all types of agents are responsible for the service executions. According to ITIL, an IT service is one provided to one or more customers by an IT organization. On the other hand, not all the instances of service providers are IT organizations, but possibly another type of agent. Therefore, the execution of an IT service occurs by means of an IT organization acting as a service provider. Nonetheless, given that IT organizations can play other roles, such as the role of a customer, it is not the case that all instances of IT organizations act as service providers in every situation, but only when they perform IT services. So, the relation between service provider and IT organization cannot be direct in any sense. On the contrary, it should be intermediated by a role (in this case, IT Service Provider - ITSP) that aggregates the criteria of identity of the species (in this case, IT Organization) and performs the mixed role (in this case, Service Provider). In this way, it is defined that not all service provider is characterized as an IT organization, which may be another type of agent. Furthermore, it is defined that there are IT organizations which in certain circumstances are service providers but not in others, possibly playing other roles. The responsibility of each played role is described by the IT service description.

The execution of an IT service occurs by means of a request, which is motivated according to the requestor's needs. Therefore, a customer is a type of requestor, i.e. an agent who requests an IT service. The process of requesting an IT service, as well as other concepts inherent to the domain of service-level management, is discussed and modeled in Costa (2008). In summary, the author describes that an IT service is appropriate for a certain need when the product from its execution satisfies the requestor's requirements, i.e. its needs. As such, an IT service can achieve a need if, and only if, the post-state of an occurrence of this service is a situation that satisfies the propositional content of the referred need, as formalized by the axiom A1.

(A1) $\forall x,y$ (IT-Service(x) $\land$ Need(y) $\land$ can-achieve(x,y) $\leftrightarrow$ $\exists a,b,c$ (ITSE(a) $\land$ Situation(b) $\land$ Proposition(c) $\land$ instance-of(a,x) $\land$ postState(b,a) $\land$ satisfies(b,c) $\land$ propositional-content-of(c,y)))

An IT service is requested by means of a document that describes the need of the requestor, termed service-level requirement (SLR). Thus, given a SLR that describes a certain need which can be achieved by an IT service, then this SLR is used to request this service (COSTA, 2008). This definition is formalized by the axiom A2.

(A2) $\forall x,y,z$ (IT-Service(x) $\land$ Need(y) $\land$ SLR(z) $\land$ can-achieve(x,y) $\land$ describes(z,y) $\rightarrow$ used-to-ask-for(z,x))

So, through a SLR, a requestor can search for an IT service that satisfies its needs. This allows the requestor to find out whether or not there are services that can achieve its needs and request, in case there is, the most adequate, according to its demand. This definition is formalized by the axiom A3.

(A3) $\forall$x,y,z,w (IT-Service(x) $\wedge$ Need(y) $\wedge$ SLR(z) $\wedge$ Requestor(w) $\wedge$ can-achieve(x,y) $\wedge$
describes(z,y) $\wedge$ used-to-ask-for(z,x) $\wedge$ characterizes(y,w) $\rightarrow$ asks-for(w,x))

Once the requester asks for services that can achieve its necessities and finds such services offered by the providers, this requestor is able to request such services. According to Costa (2008), when an IT service is requested, the requestor hires the provider responsible for this service, as formalized by the axiom A4.

(A4) $\forall$x,y,z (Requestor(x) $\wedge$ ITSP(y) $\wedge$ IT-Service(z) $\wedge$ asks-for(x,z) $\wedge$ requests(x,z) $\wedge$
provides(y,z) $\rightarrow$ hires(x,y))

According to the ITIL library, the provision of an IT service to a requestor is mediated by an agreement. Thus, once the requesting process has been established, there will be an agreement that will mediate all the IT service provision, establishing the claims and obligations related to both parts, the requester and the provider. This relation is formalized by the axiom A5.

(A5) $\forall$x,y (Requestor(x) $\wedge$ ITSP(y) $\wedge$ hires(x,y) $\rightarrow$ $\exists$z (Agreement(z) $\wedge$ mediates(z,x) $\wedge$
mediates(z,y)))

With regard to UFO, an agreement is a type of social relation. A social relation is composed of social moments, called claims and commitments. A social commitment is a type of commitment and thus the motivating cause of an action performed by an agent. Thus, the agreement established between the requestor and the provider will cause the execution of the service requested by the requestor, for it is composed of social commitments inherent to the provider.

The requesting of a service is motivated by a need of a requestor and it is fulfilled through a service execution performed by a provider. In this context, there is a relation of dependence between the requestor and the provider. From the UFO point of view, a dependence relation between agents leads to a delegation relation. According to UFO, an agent a depends on an agent b regarding a goal g if g is a goal of agent a, but a cannot achieve g and agent b can achieve g. This matter may be the reason why agent a decides to delegate such goal achievement to agent b. A delegation is thus associated with a dependency, but it is more than that. As a material relation, it is founded on more that its connected elements. In this case, the connected elements are two agents, namely, the requestor (delegator) and the provider (delegatee), as well as a goal (delegatum), which is the delegation object that represents the needs of the requestor. The foundation of this material relation is the social relator (a pair of commitments and claims), i.e. the agreement established between the two agents involved in this delegation, entitled requestor and provider. In other words, when a requestor delegates a goal to a provider, besides the fact that the requestor depends on the provider with regard to the goal, the provider commits itself to achieving the goal on behalf of the requester. This commitment is established by means of an agreement. As described in Calvi (2007), an IT service delegation (ITSD) is a delegation committed to achieve a goal according to a specific plan, the IT service. Therefore, according to UFO, an IT service delegation is a closed delegation, since it describes a specific plan which the provider should adopt to achieve the goal delegated by the requestor. In summary, if there is an agreement between the requestor and the provider which causes the execution of a service, then there will be a service delegation associated with the agreement. In addition, this delegation will be committed with the service and consequently with its execution. In this context, the requestor assumes the role of delegator while the provider assumes the role of delegatee. This relation is formalized by the axiom A6.

$$(A6)\ \forall x,y,z\ (Requestor(x) \wedge ITSP(y) \wedge Agreement(z) \wedge mediates(z,x) \wedge mediates(z,y) \rightarrow \exists w$$
$$(ITSD(w) \wedge associated\text{-}to(w,z) \wedge delegator\text{-}of(x,w) \wedge delegatee\text{-}of(y,w)))$$

As defined by UFO, when an agent is said to be able to achieve a certain goal it means that such agent can achieve this goal by itself or else delegate it to another agent that would be able to achieve it on its behalf. Thus, when a provider receives a delegation by the means of a service-level agreement, this provider analyses the delegated service and, if needed, delegates that service to other service providers, termed internal providers (e.g. an IT infrastructure department) and external providers (e.g. suppliers). In this manner, each subservice execution contributes to the service delegated by the customer, in this case, the super-service.

In view of this discussion, it is defined the relationship between business process and IT service inherent to the competency question CQ1. In summary, information technology is frequently used to support the business process activities through IT services. Given that a business process activity is an activity that is owned and performed by the business (commonly by a business unit) and an IT service is a service provided by an IT organization, the relation between business process and IT service occurs as a result of the dependence relation between the respective agents, which means business unit and IT organization. In other words, the fact of a certain business process activity occurrence being performed by the business denotes a social commitment of this agent in performing this occurrence. This social commitment contains a propositional content, i.e. a goal. Thereby this agent has a commitment to perform a certain action which satisfies this goal. In case this agent, i.e. the business, needs an IT service to achieve such a goal, this agent can delegate the goal (or part thereof) to the agent responsible for providing the IT service. In this context, the agent responsible for the business process activity assumes the role of requestor (more specifically, the role of customer) and the agent responsible for providing the service assumes the role of IT service provider. Therefore, the relation between business process and IT service is derived from the dependency relation between its respective agents, leading to a delegation relation. In this sense, it is worth noting the UFO contribution as it fundaments the entire service delegation process. Regarding the relationship between IT services and IT components, which is inherent to the competence question CQ2, the discussions during this section have defined that such components are seen as resources under the UFO point of view, assuming the usage participation. Moreover, in considering the definition of hardware and software presented by IEEE 610.10 (IEEE, 1994) as well as by ISO/IEC 2382-1 (ISO/IEC, 1993) it is possible to conclude that a hardware is a physical component that processes the instructions described by a software. In terms of UFO, a software is a type of normative description which describes a computer process, i.e. a type of event (an event universal). An instance of this process denotes an occurrence of such a process, termed computer process occurrence, i.e. a predetermined course of events whose execution includes the participation of a hardware. In this sense, the concept of hardware, as well as of software, is mapped to the notion of substantial. As such, if a service execution requires a software resource which is processed by a hardware resource, then this service execution also requires this hardware resource. In other words, if a service execution x requires a software resource y and this last resource describes a computer process z whose occurrence w includes the participation of a hardware q, then the service execution x also requires the hardware resource q, as formalized by the axiom A7.

(A7) $\forall$x,y,z,w,q (ITSE(x) $\wedge$ Software(y) $\wedge$ Computer-Process(z) $\wedge$ Computer-Process-Occurrence(w) $\wedge$ Hardware(q) $\wedge$ requires(x,y) $\wedge$ describes(y,z) $\wedge$ instance-of(w,z) $\wedge$ participation-of(w,q) $\rightarrow$ requires(x,q))

In this sense, it is useful to note the UFO contribution, as it allows the distinction between the notions pertinent to software, its processes and its occurrences, as well as the proper participations of hardware components, granting more expressiveness, clarity and veracity to the model. Indeed, the use of appropriate tools such as UFO and methods such as SABiO promoted o development of important ontological distinctions, as discussed in this section. Considering the ontology engineering approach adopted in this chapter, the next section explores the benefits from the implementation of the conceptual model.

5. An implementation and application of the proposed ontology

In view of the contributions that ontologies provide towards automated solutions, including important features such as artificial intelligence and, above all, interoperability, this section aims to present a case study in order to: (i) perform a proof of concept of the models developed in the previous section and also (ii) demonstrate how the concepts modeled by using ontologies can be implemented and applied in a computational system, in order to enable automation, including important features such as cited in this paragraph. Taking into accounting Figure 3, which illustrates the ontology development approach adopted in this work, the ontology proposed in this section concerns to the implementation model, which represents an implementation of the conceptual model proposed in the previous section.

As discussed in Guizzardi (2005, 2007), each ontology engineering phase requires the use of appropriate languages in relation to the context within which the model is being designed. In the implementation phase, the choice of a language must be conducted by the end-application requirements. This refers to languages focused on computational requirements, such as decidability and efficient automated reasoning. In terms of the configuration management process, factors such as decidability, completeness and expressiveness are considered to be key requirements because of its role in relation to all other processes in service management. Thus, for the implementation of the conceptual models, this work used the OWL DL sublanguage, since it allows a greater degree of expressiveness, as compared with OWL Lite, while maintaining computational guarantees such as completeness and decidability, features not guaranteed by OWL Full (Bechhofer et al., 2004). In addition to OWL DL, the implementation of the models also used SWRL (Horrocks et al., 2003). The SWRL language allows the representation of the axioms defined in the conceptual models presented in Section 4 in an integrated way with the concepts and relations implemented by means of OWL. Finally, for the implementation of the models in OWL and the definition of the axioms in SWRL, this work used the Protégé tool (Protégé, 2011), an ontology editor that enables the integration of different languages such as OWL and SWRL inside the same implementation environment.

Once defined the development environment, the implementation models were developed according to the modular structure of the conceptual models, as follows: (i) UFO.owl; (ii) BusinessProcess.owl; (iii) ITService.owl; (iv) ITComponents.owl and finally (v) ConfigurationItem.owl. Due to the expressivity restrictions inherent in the implementation languages, the main issue concerning the mapping from conceptual models into implementation models is related to the treatment of the reduction in semantic precision. In order to maintain this reduction at an acceptable level, the most relevant losses that were found were related to the transformation of all ontologically well-founded concepts and relations into OWL classes and properties, respectively. Regardless of the application scenario, this mapping must consider the information contained in the notation used for the development of the conceptual models, such as cardinality, transitivity, domain and range. With respect to cardinality and transitivity, in OWL it is not possible to represent them simultaneously (Bechhofer et al., 2004). As a result, this work considers that the representation of cardinality restrictions is more relevant to the implementation models developed in this section. In addition, to represent the cardinality restrictions in both directions inverse relations were used. For instance, the relation "requests" is represented by the pair of relations "requests" and "is requested by". However, according to Rector and Welty (2001), the use of inverse relations significantly increases the complexity of automated reasoning. Thus, they should be used only when necessary. With respect to domain and range, an issue that should be considered is how to organize and represent many generic relations. For example, if a generic relation "describes" is created, it is not possible to restrict the domain and the range. In this case, the design choice was to use specific relations like "describes_Software_ComputerProcess", which is represented as a sub-relation of a generic relation "describes". Finally, with respect to SWRL restrictions, this language has neither negation operators nor existential quantifiers (Horrocks et al., 2003). In addition, the SWRL language might lead to undecidable implementation models. Nevertheless, this issue may be worked around by restricting the use of rules and manipulating only those that are DL-safe (Motik et al., 2005). As an attempt to make this tangible, consider an implementation of the axiom A7, which concerns the competence question QC2, discussed in Section 4. This implementation is represented by the rule R7a.

(R7a) IT_Service_Execution(?IT-SERVICE-EXECUTION) ∧ Software(?SOFTWARE) ∧ ComputerProcess(?COMPUTER-PROCESS) ∧ ComputerProcessOccurrence(?COMPUTER-PROCESS-OCCURRENCE) ∧ Hardware(?HARDWARE) ∧ requires_ITServiceExecution_Resource(?IT-SERVICE-EXECUTION,?SOFTWARE) ∧ describes_Software_ComputerProcess(?SOFTWARE,?COMPUTER-PROCESS) ∧ isInstanceOf_ComputerProcessOccurrence_ComputerProcess(?COMPUTER-PROCESS-OCCURRENCE,?COMPUTER-PROCESS) ∧ hasParticipationOf_ComputerProcessOccurrence_Hardware(?COMPUTER-PROCESS-OCCURRENCE,?HARDWARE) → requires_ITServiceExecution_Resource(?IT-SERVICE-EXECUTION,?HARDWARE)

The axiom A7 constitutes the set of axioms that establishes the relationship between the computational resources that are required by an IT service execution as a response to the competence question QC2. Thus, this axiom involves concepts such as IT service execution, hardware and software, as well as the interrelationship between these concepts, such as the relations "requires" and "describes". As discussed earlier, concepts are implemented as classes, while relations are implemented as properties, according to OWL. According to described throughout this work, the same conceptual model can give rise to a variety of implementation models in order to meet different requirements, in accordance with the purpose of the application scenario. The rule R7a is intended to meet configuration management activities, especially the activities of identification and inference of managerial information. In Baiôco and Garcia (2010), the axiom A7 is implemented mainly in order to meet the latter activity. In this case, the axiom is implemented as formalized by the rule R7b.

(R7b) IT_Service_Execution(?IT-SERVICE-EXECUTION) $\wedge$ Software(?SOFTWARE) $\wedge$
ComputerProcess(?COMPUTER-PROCESS) $\wedge$
ComputerProcessOccurrence(?COMPUTER-PROCESS-OCCURRENCE) $\wedge$
Hardware(?HARDWARE) $\wedge$ requires_ITServiceExecution_Resource(?IT-SERVICE-
EXECUTION,?SOFTWARE) $\wedge$
describes_Software_ComputerProcess(?SOFTWARE,?COMPUTER-PROCESS) $\wedge$
isInstanceOf_ComputerProcessOccurrence_ComputerProcess(?COMPUTER-PROCESS-
OCCURRENCE,?COMPUTER-PROCESS) $\wedge$
hasParticipationOf_ComputerProcessOccurrence_Hardware(?COMPUTER-PROCESS-
OCCURRENCE,?HARDWARE) $\rightarrow$ query:select(?IT-SERVICE-EXECUTION,?SOFTWARE,
?HARDWARE) $\wedge$ query:orderByDescending(?IT-SERVICE-EXECUTION)

There are numerous contributions offered by ontology engineering for the construction of autonomous, intelligent and above all interoperable computational applications. Although done in a different area, Gonçalves et al. (2008) presents an application for the interpretation of electrocardiogram results where the use of an ontology model provides a graphical simulation of the heart behavior of an individual and the correlation of the heart behavior with the known pathologies. Regarding configuration management, this process identifies, controls, maintains and checks the versions of the existing configuration items and reports the information of the IT infrastructure to all those involved in the management. Thus, this section aims to demonstrate how implementation models can be applied in a computational environment in order to support management activities in an automated manner. In addition, the results will provide a proof of concept of the developed ontology.

The first part comprises the mapping between business processes activities and business units responsible for these activities. In addition, it includes the needs that characterize these business units. Figure 5-a shows, for example, that the business process activity BPAO_Sales is composed of the activity BPAO_Ordering, which is owned by the business unit BU_Sales, as shown in Figure 5-b. The business unit BU_Sales, in turn, has need inherent to this activity, as presented in Figure 5-c. It is worth mentioning that such information, as well as any assertion that appears highlighted in blue, concerns information previously inserted into the implementation models. On the other hand, assertions that appear highlighted in yellow are information automatically inferred by the implementation models, which denote knowledge acquisition. Such inferences are performed by the Pellet reasoner (Sirin et al., 2007).

As discussed in Section 4, IT services can achieve business needs by supporting its activities. Thus, Figure 5-d illustrates the IT services that can achieve the needs of the business. This information is inferred by executing the axiom A1, implemented in this section. Figure 5-d illustrates, for example, that the service IT_Service_Ordering can achieve the need

Need_Ordering. Figure 5-e, in turn, presents the requirements used for service requests. This inference is performed by executing the implemented axiom A2. This scenario motivates the requestor, in this case the business unit, to ask for services which can achieve its needs, as shown in Figure 5-f. This inference is performed by means of the implemented axiom A3. It should be mentioned that information of this nature is fundamental to processes such as service level management, which interacts with configuration management in requesting information in order to find services able to meet the needs of the requestors.

According to the axiom A4, once a service capable meeting the need is found, the requestor can then initiate the delegation process by contracting the service. Thus, Figure 5-g shows the provider hired by the business unit. As shown in Figure 5-g, the provider hired by the business unit BU_Sales is the IT_Department, because it is the provider responsible for providing the requested service. As discussed in Section 4, the hiring process is mediated by an agreement, as illustrated in Figure 5-h. This inference is performed by means of the implemented axiom A5. Figure 5-h illustrates, for example, that the agreement SLA_Ordering mediates the requestor BU_Sales and the provider IT_Department. This agreement, in turn, characterizes the delegation process, which has a delegator (in this case the requestor) and a delegatee (in this case the provider), as shown in Figure 5-i. This inference is performed by means of the implemented axiom A6. Figure 5-i illustrates, for example, that the delegation ITSD_Ordering is associated to the agreement SLA_Ordering and has as the delegator the BU_Sales and as the delegatee the IT_Department.

As described in Section 4, the hired provider receives the service delegation from the requestor and provisions the necessary resources for the service execution. In this sense, the received service execution is characterized as a complex action which is delegated to the support groups by means of subservices. In this context, the hired provider, in a manner similar to that of the business unit, plays the role of requestor and the support groups, in turn, play the role of service provider. Thus, Figure 5-j shows, for instance, that the execution ITSE_Ordering, which represents an instance of the service IT_Service_Ordering, is composed of the sub-executions ITSE_Ordering_Processing and ITSE_Ordering_Printing. The delegation performed by the provider to the support groups is mediated by agreements. Figure 5-k shows, for example, that the agreement OLA_Ordering_Processing mediates the delegation process of the sub-execution ITSE_Ordering_Processing between the IT_Department and the IT_Department_System. In this case, IT_Department plays the role of delegator while the support groups play the role of delegatee, as shown in Figure 5-l.

IT services are based on the use of information technology. Thus, Figure 5-m relates the software required by each service execution. Figure 5-m shows, for example, that the execution ITSE_Ordering_Processing requires the software Software_Ordering_Processing. As discussed in Section 4, software is processed by hardware. Thus, Figure 5-n presents the hardware associated with the processing of the concerned software. In addition, Section 4 states that if a service execution requires a software and this software is processed by a hardware, then this service execution also requires this hardware, as illustrated in Figure 5-o. This inference is performed by means of the implemented axiom A7. Figure 5-o illustrates, for example, that the execution ITSE_Ordering_Processing requires the hardware Hardware_Sales, since such hardware processes the software Software_Ordering_Processing (as shown in Figure 5-n) required by such an execution.

Thus, this case study concludes the mapping between business and IT. This mapping is fundamental to other management processes. As an illustration, the configuration management process of this case study is able to correlate and determine that a particular event of unavailability on the hardware Hardware_Sales affects the software Software_Ordering_Processing, which is used by the service execution ITSE_Ordering_Processing. This execution is part of the execution ITSE_Ordering which is instance of the service IT_Service_Ordering and, in turn, supports related activity of the business process activity BPAO_Sales. This correlation, provided by the implementation model, is the basis for activities such as: (i) event correlation in event management; (ii) workaround identification in incident management; (iii) root cause analysis in problem management and (iv) impact analysis in change management.

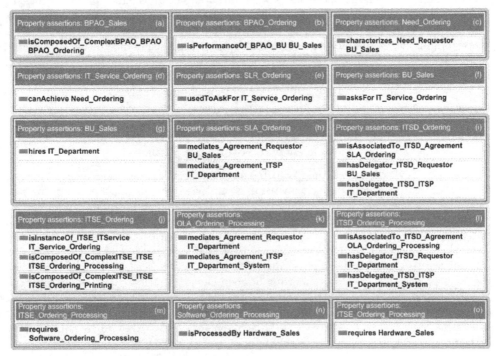

Fig. 5. Application of the implementation model.

To complete this case study, consider intelligent software agents playing the roles of requestor and provider and, consequently, negotiating the provision of services that meet the needs of the environment. Regarding the role of provider, such activities refer to the various management disciplines, as presented throughout this section. In particular, this scenario denotes implementation models subsidizing paradigms known as autonomous networks. In general, it denotes implementation models promoting automation in various areas of interest.

6. Conclusion

As discussed in this chapter, automation enables organizations to explore opportunities as well as supporting challenges in an effective and efficient way. The important role played by

IT as an instrument for automation has made automation increasingly dependent on IT, consequently rising the demands for advances, as observed by the growing challenges arising from the conception of systems continuously more complex, intelligent and, above all, interoperable. Moreover, the need for an efficient and effective IT management has grown substantially, as evidenced by widespread adoption of innovative best practices libraries and standards. For this reason, this work presented an ontology of IT service configuration management. The objective was not only to adopt the state of the art in order to address key research challenges in IT management, but also to foment novel approaches which can be applied in IT in various areas of interest.

The diverse uses attributed to the ontologies in computer science and the interrelation between their purposes promote the search for approaches capable of providing the construction of ontological models able to achieve the various objectives assigned to them. Based on innovative and high quality research initiatives, this chapter discussed about a systematic approach for building ontologies known as Ontology Engineering. In considering the various uses and purposes, as well as their interrelationships, these initiatives attempt to establish a structured means of development as an alternative to the various *ad hoc* approaches that characterize the current developments and imply in models unable to achieve their goals. In summary, this approach allowed the development of conceptual models which are application-independent artifacts and, as a result, it enabled their use as a reference ontology for the subsequent development phases, deriving implementation models in order to address the different purposes of end applications.

According to Guizzardi and Halpin (2008), the practice of conceptual modeling is permeated by philosophical questions. This demonstrates the need for an appropriate theoretical foundation for conceptual modeling languages so as to ensure that the quality requirements of domain and comprehensibility appropriateness can be fulfilled by the produced conceptual models. In this sense, they advance that philosophically well-founded ontologies play a key role in this initiative. They complement this line of reasoning by citing Guarino and Guizzardi (2006) and emphasizing that although typical conceptual modeling languages provide facilities for structuring domain elements, such as taxonomies and data value structures, the justification for the validity of many structuring choices, as much as the justification for the grammar of many natural language sentences, can only be made on ontological grounds, in this sense, on a philosophical basis. As a final consideration, Guizzardi and Halpin (2008) point out that philosophical foundations are vital components with respect to conceptual modeling, in general, and domain ontology engineering, in particular, as mature disciplines with sound principles and practices. Thus, in quoting the physicist and philosopher of science Mario Bunge, "every science presupposes some metaphysics", they conclude that a scientific field can either choose to develop and make explicit its philosophical foundations or to remain oblivious to its inevitable and often *ad hoc* ontological and epistemological commitments.

Accordingly, in addition to the appropriate methods and techniques, such as the SABiO method, this approach used a philosophically well-founded ontology, termed UFO. The SABiO method provided a systematic approach that led the development of ontology proposed in this chapter, describing an iterative process, closely related to evaluation. Thus, with emphasis on the concept of competence questions, the SABiO method provided a means for defining the scope and purpose of the ontology, leading its capture and formalization, the reuse of existing ontologies, as well as its evaluation and documentation, enabling the ontology proposed in this chapter to adequately meet the requirements for which it was designed, as confirmed by proof of concept. The UFO ontology, in turn, was useful in building a conceptual model committed to maximizing the expressivity, clarity and truthfulness of the modeled domain. These characteristics are key quality attributes of a conceptual model, responsible for its effectiveness as a reference framework for the tasks of semantic interoperability and reuse. In fact, as a knowledge representation language philosophically well-founded, the UFO ontology aims to provide a sound basis for the representation of a conceptualization and therefore inhibit arbitrary descriptions of concepts and relationships of a universe of discourse. As discussed during the development of the conceptual models proposed in this chapter, the UFO ontology guided diverse modeling decisions, contributing to the derivation of new knowledge or the identification and elucidation of ambiguous and inconsistent representations of the domain, often represented in various literatures.

As discussed in this chapter, the more it is known about a universe of discourse and the more precisely it is represented, the bigger the chance of producing models that reflect, as much as possible, the appropriate conceptualization of the domain. In this sense, besides the use of appropriate languages, such as ontologies philosophically well-founded, as a common and shared specification, it is important that the specification of the domain considers appropriate literatures and, especially, that it considers its main concepts and relationships as well as application independence. Considering the universe of study of this chapter, these aspects associated with the use of appropriate methods and tools enabled the development of models able to maximize the alignment between IT and business for humans and computers. Moreover, these aspects allude to a point of view which should be mentioned. By maximizing the capacity of a model in acting as a common and shared source about a universe of discourse, conceptual models, representing norms and standards, can be potentially used as an addendum to such literatures. This is because, in general, these libraries are described in natural languages, which are susceptible to ambiguities and inconsistencies, as opposed to conceptual models, which are formally described.

Considering the importance of automation, as well as the contributions that an implementation provides in terms of ontology evaluation, it was developed an implementation model, derived from the conceptual model proposed in this chapter. In addition, by applying the entire approach discussed in this chapter, it is possible to attest it, as well as making it more tangible, promoting its benefits. It should be mentioned that the development of conceptual models followed by the development of implementation models became evident the distinction between ontology representation languages, as discussed throughout this chapter. This demonstrates that the approach adopted in this chapter shows itself appropriate by considering the various uses and purposes assigned to ontologies. In this way, despite the expressivity restrictions inherent in implementation languages, it was possible to perform a proof of concept of the ontology developed in this chapter as well as demonstrating how such models can be derived and implemented in computing environments with a view to the different computational requirements. In particular, it was possible to show how implementation models can support automation, including special characteristics such as knowledge acquisition and interoperability. From this point of view it is important to note that the concepts inherited from UFO are important in promoting paradigms such as artificial intelligence by making explicit, for computational agents, concepts that express the daily lives of human agents, such as intentions, goals and actions. These factors are especially important in face of the increasing need for integration of complex as well as heterogeneous systems and also when considering autonomous systems.

Therefore, the contributions of this chapter are not restricted to the domain of configuration management. Instead, they promote semantic interoperability in IT in diverse areas of interest, maximizing the advances in automation. Additionally, this chapter makes possible other researches. In this sense, future works include: (i) the extension of the ontology for covering other business-driven IT management concepts, such as IT services metrics and business measures as well as their relationships, improving the alignment between these two domains; (ii) the extension of the ontology in order to cover other configuration management concepts, such as baseline, version and variant; (iii) the extension of the ontology to cover other management process, such as change management and release management; (iv) the application of conceptual models as an addendum to norms and standards; (v) the application of the implementation models, especially with techniques like artificial intelligence, promoting paradigms such as autonomous networks.

7. References

Baiôco, G., Costa, A.C.M., Calvi, C.Z. & Garcia, A.S. (2009). IT Service Management and Governance – Modeling an ITSM Configuration Process: a Foundational Ontology Approach, In *4th IFIP/IEEE International Workshop on Business-driven IT Management, 11th IFIP/IEEE International Symposium on Integrated Network Management*, 2009

Baiôco, G. & Garcia, A.S. (2010). Implementation and Application of a Well-Founded Configuration Management Ontology, In *5th IFIP/IEEE International Workshop on Business-driven IT Management (BDIM), 12th IFIP/IEEE Network Operations and Management Symposium (NOMS)*, Osaka, 2010

Bechhofer, S. et al. (2004). OWL Web Ontology Language Reference, In *W3C Recommendation*, Oct 2011, Available from http://www.w3.org/TR/owl-ref/

Brenner, M., Sailer, M., Schaaf, T. & Garschhammer, M. (2006). CMDB - Yet Another MIB? On Reusing Management Model Concepts in ITIL Configuration Management, In *17th IFIP/IEEE Distributed Systems Operations and Management (DSOM)*, 2006

Calvi, C. Z. (2007). IT Service Management and ITIL Configuration Process Modeling in a Context-Aware Service Platform (in Portuguese), Master Dissertation, UFES, 2007

Costa, A.C.M. (2008) ITIL Service Level Management Process Modeling: An Approach Using Foundational Ontologies and its Application in Infraware Platform (in Portuguese), Master Dissertation, UFES, 2008

Falbo, R.A. (1998). Knowledge Integration in a Software Development Environment (in Portuguese), Doctoral Thesis, COPPE/UFRJ, 1998

Falbo, R.A. (2004). Experiences in Using a Method for Building Domain Ontologies, In *16th Conference on Software Engineering and Knowledge Engineering (SEKE)*, Canada, 2004

Gonçalves, B. et al. (2009). An Ontology-based Application in Heart Electrophysiology: Representation, Reasoning and Visualization on the Web, In *ACM Symposium on Applied Computing*, 2009

Gonçalves, B.N., Guizzardi, G. & Pereira Filho, J.G. (2011). Using an ECG reference ontology for semantic interoperability of ECG data, In *Journal of Biomedical Informatics*, Special Issue on Ontologies for Clinical and Translational Research, Elsevier, 2011

Guarino, N., Guizzardi, G. (2006). In the Defense of Ontological Foundations for Conceptual Modeling, In *Scandinavian Journal of Information Systems*, ISSN 0905-0167, 2006

Guizzardi, G. & Wagner, G. (2004). A Unified Foundational Ontology and some Applications of it in Business Modeling, In *Open INTEROP Workshop on Enterprise Modelling and Ontologies for Interoperability, 16th CAiSE*, Latvia, 2004

Guizzardi, G. (2005). *Ontological Foundations for Structural Conceptual Models*, Ph.D. Thesis, University of Twente, ISBN 90-75176-81-3, The Netherlands, 2005

Guizzardi, G. (2006). The Role of Foundational Ontology for Conceptual Modeling and Domain Ontology Representation, In *Companion Paper for the Invited Keynote Speech, 7th International Baltic Conference on Databases and Information Systems*, 2006

Guizzardi, G. (2007). On Ontology, ontologies, Conceptualizations, Modeling Languages, and (Meta)Models, In *Frontiers in Artificial Intelligence and Applications, Databases and Information Systems IV*, IOS Press, ISBN 978-1-58603-640-8, Amsterdam, 2007

Guizzardi, G., Falbo, R.A. & Guizzardi, R.S.S. (2008). The Importance of Foundational Ontologies for Domain Engineering: The case of the Software Process Domain (in Portuguese), *IEEE Latin America Transactions*, Vol. 6, No. 3, July 2008

Guizzardi, G. & Halpin, T. (2008). Ontological foundations for conceptual modeling, In *Journal of Applied Ontology*, v.3, p.1-12, 2008

Guizzardi, G., Lopes, M., Baião, F. & Falbo, R. (2009). On the importance of Truly Ontological Distinctions for Ontology Representation Languages: An Industrial Case Study in the Domain of Oil and Gas, In *Lecture Notes in Business Information Processing*, 2009

Horrocks, I. et al. (2003). SWRL: A Semantic Web Rule Language Combining OWL and RuleML, In *DAML*, Oct 2011, From http://www.daml.org/2003/11/swrl/

IEEE (1994). *IEEE Standard Glossary of Computer Hardware Terminology*, IEEE Std 610.10-1994

ISO/IEC (1993). *IT – Vocabulary – Part 1: Fundamental terms*, ISO/IEC 2382-1:1993

ISO/IEC (2005). *Information technology – Service management*, ISO/IEC 20000, 2005

ITIL (2007). *ITIL Core Books*, Office of Government Commerce (OGC), TSO, UK, 2007

Jones, D.M., Bench-Capon, T.J.M. & Visser, P.R.S. (1998). Methodologies For Ontology Development, In *15th IFIP World Computer Congress*, Chapman-Hall, 1998

Lopez de Vergara, J.E., Villagra, V.A. & Berrocal, J. (2004). Applying the Web Ontology Language to management information definitions, *IEEE Communications Magazine*

Majewska, M., Kryza, B. & Kitowski, J. (2007). Translation of Common Information Model to Web Ontology Language, In *International Conference on Computational Science*, 2007

Moura, A., Sauve, J. & Bartolini, C. (2007). Research Challenges of Business Driven IT Management, In *2nd IEEE/IFIP Business-driven IT Management, 10th IFIP/IEEE IM*

Moura, A., Sauve, J. & Bartolini, C. (2008). Business-Driven IT Management - Upping the Ante of IT: Exploring the Linkage between IT and Business to Improve Both IT and Business Results, In *IEEE Communications Magazine*, vol. 46, issue 10, October 2008

Motik, B., Sattler, U. & Studer, R. (2005). Query Answering for OWL-DL with Rules, In *Journal of Web Semantics*, Vol. 3, No. 1, pp. 41-60, 2005

Pavlou, G. & Pras, A. (2008). Topics in Network and Service Management, In *IEEE Communications Magazine*, 2008

Pras, A., Schönwälder, J., Burgess, M., Festor, O., Pérez, G.M., Stadler, R. & Stiller, B. (2007). Key Research Challenges in Network Management, *IEEE Commun Magazine*, 2007

Protégé (2011). The Protégé Ontology Editor and Knowledge Aquisition System, October 2011, Available from http://protege.stanford.edu/

Rector, A. & Welty C. (2001). Simple part-whole relations in OWL Ontologies, In *W3C Recommendation*, Oct 2011, Available from http://www.w3.org/2001/sw/BestPractices/OEP/SimplePartWhole

Sallé, M. (2004). IT Service Management and IT Governance: Review, Comparative Analysis and their Impact on Utility Computing, In *HPL-2004-98*, 2004

Santos, B.C.L. (2007). O-bCNMS: An Ontology-based Network Configuration Management System (in Portuguese), Master Dissertation, UFES, 2007

Sirin E. et al. (2007). Pellet: a Practical OWL-DL Reasoner, In *Journal of Web Semantics: Science, Services and Agents on the World Wide Web*, Vol. 5, No. 2, pp 51-53, 2007

Smith, B. & Welty, C. (Eds.). (2001). Ontology: Towards a new synthesis, Chris Welty and Barry Smith, Formal Ontology in Information Systems, ACM Press, 2001

Ullmann, S. (1972). Semantics: An Introduction to the Science of Meaning, Basil Blackwell, Oxford, 1972

Vermeer, M.W.W. (1997). Semantic interoperability for legacy databases, PhD Thesis, University of Twente, The Netherlands, 1997

Wong, A.K.Y., Ray, P., Parameswaran, N. & Strassner, J. (2005). Ontology Mapping for the Interoperability Problem in Network Management, *IEEE Journal on Selected Areas in Communications (JSAC)*, Vol. 23, No. 10, October 2005

Xu, H. & Xiao, D. (2006). A Common Ontology-Based Intelligent Configuration Management Model for IP Network Devices, *In Proceedings of the First International Conference on Innovative Computing, Information and Control (ICICIC)*, 2006

3

Automation of Subjective Measurements of Logatom Intelligibility in Classrooms

Stefan Brachmanski
Wroclaw University of Technology
Poland

1. Introduction

A great number of rooms are dedicated to a voice communication between a singular speaker and a group of listeners. Those rooms could be as small as meeting room and classrooms or larger like auditoriums and theatres. There is a major demand for them to assure the highest possible speech intelligibility for all listeners in the room. Classrooms are an example of rooms where a very good speech intelligibility is required (a teacher talks to a group of students who want to understand the teacher's utterance). To determine the intelligibility degree (its maps) of rooms, it is necessary to take measurements in many points of those rooms. The number of measurement points depends on room's size and precision of created intelligibility room's maps. Despite of the crucial progress in the instrumental measurement techniques, the only reliable method subjective speech intelligibility measurement is still very time consuming, expensive, demanding high skills and specially trained group of listeners. The first part of this chapter presents the idea of speech intelligibility subjective measurements. The measurements with properly trained team are taken in described standards, conditions have to be controlled and repeated. The subsequent sections of the chapter are focusing on one of the classic subjective speech intelligibility measurement method in rooms (classrooms) and its automated version which is named as the modified intelligibility test with forced choice (MIT-FC). Finally, in the last section of the chapter, there compared results of subjective speech intelligibility measurements in rooms taken with classic and automated methods are and the relation between intelligibility taken with the forced choice method is given as well. The presented relation let us compare results taken with both methods and use relations known from earlier research carried in domain of speech intelligibility. However, the biggest advantage of the speech intelligibility measurement automation is the shortening of measurement time and the possibility of taking simultaneous measurements in several points of the room. The result of speech intelligibility is obtained just after the end of measurements, it is then possible to obtain the intelligibility map in few minutes. Of course the precision is growing with number of listeners in particular points.

Speech quality is a multi-dimensional term and a complex psycho-acoustic phenomenon within the process of human perception. Every person interprets speech quality in a different way. The pioneering work on speech intelligibility was carried out by Fletcher at Bell Labs in the early 1940s. Fletcher and his team not only established the effects of bandwidth on intelligibility but also the degree to which each octave and ⅓-octave band contributed.

One of the fundamental parameters for quality assessment of speech signal transmission in analogue and digital telecommunication chains as well as in rooms, sound reinforcement systems and at selection of aural devices is speech intelligibility (ANSI, 2009; Basciuk & Brachmanski, 1997; Brachmanski, 2002, 2004; Davies, 1989; International Organization for Standardization – [ISO], 1991; Majewski, 1988, 1998, 2000; Polish Standard, 1991, 1999; Sotschek 1976). Satisfactory speech intelligibility should be provided by telecommunication channels, rooms and hearing aids. It is obvious, that speech intelligibility concerns only the linguistic information (i.e. what was said) and does not take into account such features of speech, like its naturalness or the speaker voice individuality. Nevertheless, intelligibility should be and - up till now - is viewed as a basic and most important aspect of the quality of all systems which transmit, code, enhance and process the speech signals. Satisfactory speech intelligibility requires adequate audibility and clarity.

In general the evaluation of the speech quality may be done by subjective (intelligibility, quality rating) (Farina, 2001; Howard & Angus, 2009; Möller, 2010) and objective methods e.g. Articulation Index - AI (American National Standards Institute – [ANSI], 1997; French & Steinberg, 1947; Kryter, 1962), Speech Transmission Index - STI (Brachmanski, 1982, 2004, 2006, Houtgast & Steeneken, 1973; International Electrotechnical Commission [IEC], 1991; Lam & Hongisto, 2006; Steeneken & Houtgast, 1980, 2002), Perceptual Evaluation of Speech Quality – PESQ (International Telecommunication Union - Telecommunication Standardization Sector [ITU-T], 2003), Perceptual Objective Listening Quality Assessment – POLQA, (ITU-T, 2011) . Designers of devices and systems intended for speech transmission incline to usage of objective measurement methods, not always taking into consideration the application limitations and exactitude which depends on type of examined object and measure conditions. However the final verifier of quality of speech transmission devices is their user, that is to say man.

The aim of aural evaluation is quantitative evaluating and qualitative differentiation of acoustical signals reaching a listener. The process of aural evaluation can be presented as follows

$$B \rightarrow S \rightarrow R \tag{1}$$

where: B – stimulus reaching the listener, S – listener, R – listener's reaction

The reaction R is dependent on the signal S reaching the listener's receptors and on conditions in which the listener is based. Generally it can be assumed that the reaction R depends on the sum of external stimuli having the effect on the listener and internal factors having the effect on his organism. That assumption, however, doesn't take into consideration the listener's characteristic features such as cognitive abilities, rate of information processing, memory etc. The reaction can be then presented as a function

$$R = f(B,S) \tag{2}$$

Physiological and psychological process connected with a reaction to sound signal (audio) consists of sensational reaction and emotional reaction. The total listener's reaction is a sum of both types of reaction.

The sensational reaction is the effect of a physiological process which occurs during the listening. It arises when a certain stimulus overdraws sensitivity levels or aural sensation category levels. The emotional reaction is more complex and difficult in analysis because it isn't a direct result of received signal features but the listener's habits and individuality. As a conclusion it can be stated that the sensational reaction is the reflection of an acoustical picture created in a person's (listener's) mind, whereas the emotional reaction is the reflection of a person's attitude to that picture. Psychological and psychoacoustical research have proven that when provided stable in time conditions of evaluation, the differences in sensational reactions of particular listeners are substantially smaller than the differences in their emotional reactions. Therefore, one aim of objectivisation of aural evaluation is limiting the influence of emotional reaction on the final assessment result. That aim is achieved by introducing appropriately numerous assessments statistics, a proper choice and training of a listeners team and proper choice of testing material and rules of carrying listeners tests. The results analysis also has the big role in minimization the influence of the emotional reaction on the assessment result.

Among the different subjective methods that have been proposed for assessment of speech quality in rooms, the preferred are methods based on intelligibility tests or listening-only tests (ITU-T, 1996a, 1996b; ITU-R, 1997). The subjective measure results should be mostly dependent on physical parameters of the tested room and not on the structure of the tested language material. The elimination of semantic information is done by means of logatom[1] (i.e. pseudo-words) lists on the basis of which the logatom intelligibilities are obtained. The problem ist hat speech (logatom) intelligibility is not a simply parameter to measure.

2. The traditional method of logatom intelligibility

The measurement of logatom intelligibility consists in the transmission of logatom lists, read out by a speaker, through the tested channel (room), which are then written down by listeners and the correctness of the record is checked by a group of experts who calculate the average logatom intelligibility. It is recommended to use lists of 50 or 100 logatoms (Fig.1).

Logatom lists are based on short nonsense word of the CVC type (consonant-vowel-consonant). Sometimes only CV, but also CVVC, CVCVC-words are used. The logatoms are presented in isolation or in carrier phrase e.g. "Now, please write down the logatom you hear". Each list should be phonetically and structurally balanced (Fig.2).

---

[1] Logatom – (*logos* (gr) - spoken phrase, *atom* (gr) – indivisible) vocal sound, generally insignificant, usually made by the sound of a consonant or the first consonant, then by an intermediate vowel, finally by a consonant or a final consonant sound.

LOGATOM LIST 3.1

1 - 10	11 - 20	21 - 30	31 - 40	41 - 50
dufcze	cze	teń	jo	czalmy
ajtes	wać	foli	es	ny
wesk	niacko	delak	gapysz	nij
pi	wespa	szo	tyr	dnaf
żewan	szkio	penia	geja	uzy
pracło	łare	si	lo	nio
nujec	le	weru	ju	towir
wajt	wnieśma	krwolta	dziorka	ła
be	myśro	ponta	jeskol	woszki
szonie	ri	tyn	ca	krynom

51 - 60	61 - 70	71 - 80	81 - 90	91 - 100
chfypa	dy	pa	gruto	słynej
ca	zjaciech	piestma	ke	jentuś
czniesa	wyzo	fa	todzi	jos
szaźmo	żam	żnota	ply	bep
neni	fjesztech	gzał	abe	du
se	ro	dojsto	tnązy	wo
błunki	kiet	ga	kamu	cioch
dnozger	desi	vrum	ksi	zał
bomuć	mo	iżo	wnici	ną
dzany	bejce	du	nu	sze

Fig. 1. Illustrative 100-logatom list from set 3 for the Polish language.

The measurement should be carried out in rooms in which level of internal noise together with external noise (not introduced on purpose) does not exceed 40 dBA. If no requirements as to background noise are specified for a tested chain, articulation should be measured at a noise level of 60 dBA in the receiving room and for the Hoth spectrum (Fig. 3).

The listeners should be selected from persons who have normal, good hearing and normal experience in pronunciation in the language used in the test. A person is considered to have normal hearing if her/his threshold does not exceed 10 dB for any frequency in a band of 125 Hz –4000 Hz and 15 dB in a band of 4000 Hz– 6000 Hz. Hearing threshold should be tested by means of a diagnostic audiometer.

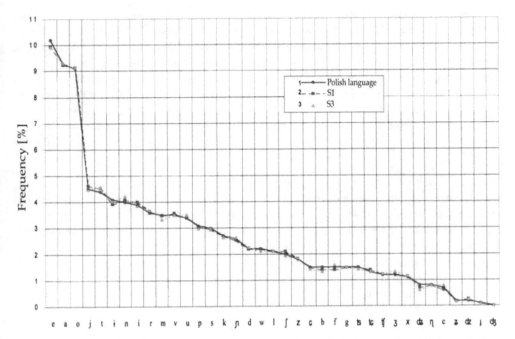

Fig. 2. Example of phonetic balance of two sets, three lists of 100 logatom each for the Polish language.

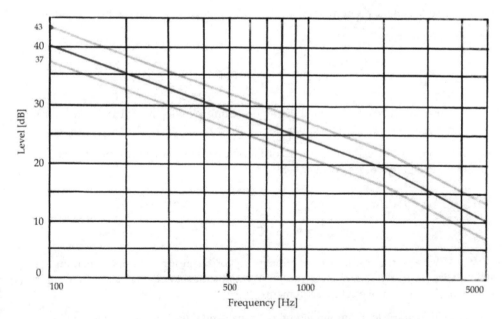

Fig. 3. The room noise power density spectrum (Hoth noise) (Polish Standard 1991, 1999).

The size of the listening group should be such that the obtained averaged test results do not change as the group size is further increased (minimum 5 persons). The group of listeners who are to take part in logatom intelligibility measurements should be trained (2-3 training sessions are recommended). Logatoms should be spoken clearly and equally loudly without accenting their beginnings or ends. The time interval between individual logatoms should allow the listener to record the received logatom at leisure. It is recommended that logatoms should be spoken with 3-5 sec. pauses in between. The time interval between sessions should not be shorter than 24 h and not longer than 3 days. The total duration of a session should not exceed 3 hours (including 10 minute breaks after each 20 minute listening period).

Listeners write the received logatoms on a special form on which also the date of the test, the test list number, the speaker's name or symbol (no.), the listener's name and additional information which the measurement manager may need from the listener is noted. The recording should be legible to prevent a wrong interpretation of the logatom. The received logatoms may be written in phonetic transcription (a group of specially trained listeners is needed for this) or in an orthographic form specific for a given language. In the next step, the group of experts checks the correctness of received logatoms and the average logatom intelligibility is calculated in accordance to the equation (3) and (4)

$$W_L = \frac{1}{N \cdot K} \sum_{n=1}^{N} \sum_{k=1}^{K} W_{n,k} \ [\%] \tag{3}$$

N - number of listeners, K - number of test lists, $W_{n,k}$ - logatom intelligibility for n-th listener and k-th logatom list,

$$W_{n,k} = \frac{P_{n,k}}{T_k} \cdot 100 \, [\%] \tag{4}$$

$P_{n,k}$ - number of correctly received logatoms from k-th logatom list by n-th listener, T_k - number of logatoms in k-th logatom list.

Standard deviation s, calculated in accordance to Eq.(3), expresses the distribution of logatom intelligibility values W_L over listeners.

$$s = \left[\frac{1}{N \cdot K - 1} \sum_{n=1}^{N} \sum_{k=1}^{K} (W_{n,k} - W_L)^2 \right]^{1/2} \tag{5}$$

If $|W_{n,k} - W_L| > 3s$, the result of measurement is not taken into account, when an average intelligibility is calculated and calculation of W_L and s must be done in accordance to Eq. (1) and (2) for reduced number of measurements.

The obtained average logatom intelligibility value can be used to determine quality classes according to Table 1 (Polish Standard 1991).

Quality class	Description of quality class	Logatom intelligibility [%]
I	Understanding transmitted speech without slightest concentration of attention and without subjectively detectable distortions of speech signal	above 75
II	Understanding transmitted speech without difficulty but with subjectively detectable distortions of speech	60 ÷ 75
III	Understanding transmitted speech with concentration of attention but without repetitions and return queries	48 ÷ 60
IV	Understanding transmitted speech with great concentration of attention and with repetitions and return queries	25 ÷ 48
V	It is impossible to fully understand transmitted speech (breakdown of communication)	to 25
For each quality class lowest logatom articulation values are lowest admissible values		

Table 1. Speech intelligibility quality classes for analog channels in the traditional logatom intelligibility method.

3. Modified intelligibility test with forced choice (MIT-FC)

In the traditional intelligibility tests the listeners write down (in ortographic form) received utterences on a sheet of paper and next a professional team revises the results and calculates the average intelligibility. This is the most time-consuming and difficult operation. To eliminate "hand-made " revision of the tests a method, called *"modified intelligibility test with forced choice"* (MIT-FC) has been designed and investigated in the Institute of Telecommunications, Teleinformatics and Acoustics at the Wroclaw University of Technology .

In the MIT-FC method all experiments are controlled by a computer. The automation of the subjective measurement is connected with the basic change in generation of logatoms and in making decision by a listener. The computer generates logatoms and presents the utterances (for logatom test the list consists of 100 phonetically balanced nonsense words), via a D/A converter and loudspeaker to the listeners subsequently and for each spoken utterance several logatoms that have been previously selected as perceptually similar are visually presented.

It has been found (Brachmanski, 1995) that the optimal number of logatoms presented visually to the listeners is seven (six alternative logatoms and one transmitted logatom to be recognized). The listener chooses one logatom from the list visually presented on the computer monitor. The computer counts the correct answers and calculates the average logatom intelligibility and standard deviation. The measurement time for one logatom set (3 lists) consisted of 300 logatoms is 20 minutes.

All measurement procedures are fully automatized and an operator has a flexible possibilities to set the measurement parameters and options. It is also possible to upgrade the application which realizes the MIT-FC method with more sophisticated scores processing. Block diagram system for the subjective measurements of logatom intelligibility by MIT-FC method in the rooms is presented in Fig.4.

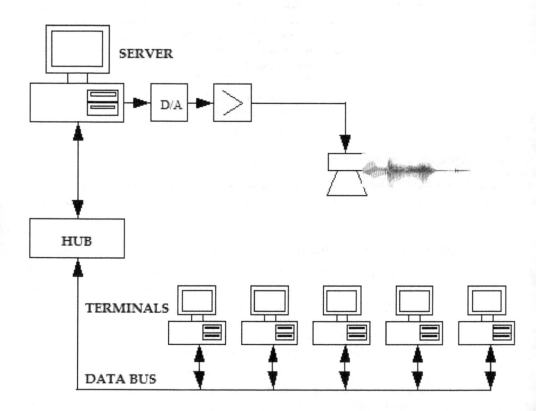

Fig. 4. The measuring system for the assessment of logatom intelligibility in room.

The obtained with MIT-FC method average logatom intelligibility value can be used to determine quality classes according to Table 2.

Quality class	Description of quality class	Logatom intelligibility [%]
I	Understanding transmitted speech without slightest concentration of attention and without subjectively detectable distortions of speech signal	above 70
II	Understanding transmitted speech without difficulty but with subjectively detectable distortions of speech	60 ÷ 70
III	Understanding transmitted speech with concentration of attention but without repetitions and return queries	54 ÷ 60
IV	Understanding transmitted speech with great concentration of attention and with repetitions and return queries	40 ÷ 54
V	It is impossible to fully understand transmitted speech (breakdown of communication)	to 40
For each quality class lowest logatom articulation values are lowest admissible values		

Table 2. Classes of speech intelligibility quality for analog channels for the MIT-FC method.

4. MIT-FC measurement system

The program for the subjective assessment of speech transmission in rooms with logatom intelligibility method with forced choice (MIT-FC) is based on TCP Client/Server technology i.e. the communication is done by local network. Requirements of the program are following: PC computer with Windows 9x, a network card and hub for communication between the Server and Clients (members of the team of listeners).

The work with the program starts with the installation on the Server computer a Server program. The Server is supervised by the person leading the subjective assessment. The next step is the installation of the Client program on the Client (listener) computers. The Client computers are used by members of the team of listeners. Before starting the assessment the Server and Client programs should be configured. During the configuration of the Server program it is necessary to give the path to the directory with signal files (testing signals - logatoms), number of logatoms per a session, intervals between reproduced logatoms, number of sessions and the port for the communication with the Client (usually 3000) (Fig.5).

The configuration of the Client program is done by giving the name of the measure point (e.g. a room, location of listener in the room etc.), listeners login, IP address and the port number (usually: 3000) (Fig.6).

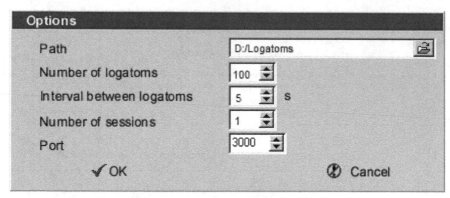

Fig. 5. The example of the window of the Server program

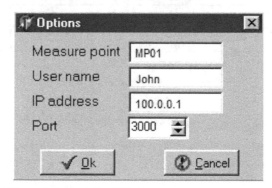

Fig. 6. The example of the window of the Client program

The next step after the configuration is the connection process. The number of people being logged on can be seen on the right-hand side of the main window of the program. During a multi-session assessment (the number of session should be chosen during the configuration of the Server), after finishing of the first session the program waits for reconfiguration of measurement positions. For example, changing the listener on certain measurement position, we should first disconnect the Client and change the user name. After this and choosing the Continue option, the next session can be started. The computer generates logatoms and for each spoken utterance visually shows six alternative logatoms and one transmitted logatom to be recognized (Fig.7a). The listener chooses one logatom from the list visually presented on the computer monitor (Fig. 7b). During the tests, the listener is confirming his response by using key from '1' to '7'. Other keys are non-active during all testing session.

After finishing all measurement sessions, the dialog window presenting results in two options shows up:

1. Result of session nr - in this option the number of session for which the results will be shown should be given.
2. Summary - in this option the summary of all sessions with detailed list of listeners and measurement points shows up.

a) 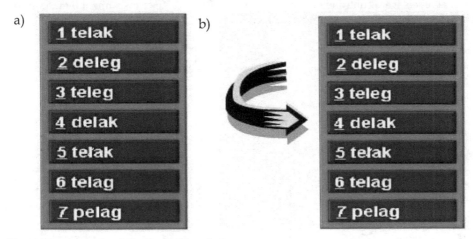 b)

Fig. 7. The example of the listener's window.

5. Experiments

The goals of experiment

- decision if the results of traditional and modified with forced choice methods let finding the relation which would allow to convert results from one method to the other and the classification of rooms tested with both methods,
- measurement of experimental relations between traditional and modified logatom intelligibility methods.

Taking into account the comparative character of the experiment it was planned to be done only in a function of used logatom intelligibility method. With this end in view each measurement was done with both traditional and modified methods, not changing:

- listeners,
- surroundings.
- measurement system (only the logatom lists),

The subjective tests were done according to Polish Standard PN-90/T-05100   with the team of listeners made up of 12 listeners in age from 18 to 25 years. The listening team was selected from students at Wroclaw University of Technology with normal hearing. The qualification was based on audiometric tests of hearing threshold. The measurements of logatom intelligibility were done using the traditional method and the MIT-FC method. The measurements were taken in two unoccupied rooms (Fig. 8). In each room, four measure point (Mp) were selected. These positions were chosen in the expectation of yielding a wide range of logatom intelligibility. Sound sources (voice and white or rose noise) were positioned in the part of the room normally used for speaking. One loudspeaker was the voice source and the second – the noise source. The various conditions were obtained by combination 14 level of white and rose noise and four measure point. The 14 signal-to-noise ratios (SNR) were used: 39 (without noise, only background noise), 36, 33, 30, 27, 24, 21, 18, 15, 12, 9, 6, 3, 0dB. As a result

112 different transmission conditions were obtained (14 SNR*4Mp*2rooms). The speech and noise signal levels were controlled by means of a 2606 Bruel & Kjaer instrument by measuring them on a logarithmic scale according to correction curve A.

Fig. 8. Plan view of the rooms 1 and 2 showing source position and four receiver positions (**I** – loudspeaker – source of the logatoms, **III** - loudspeaker – source of the noise, **1, 2, 3, 4** – measurement points).

The testing material consisted of phonetically and structurally balanced logatoms and sentences lists uttered by professional male speaker, whose native language was Polish. Logatom lists reproduced through the loudspeaker were recorded on the digital tape recorder in an anechoic chamber using a linear omnidirectional microphone. The microphone was positioned 200mm from the speaker's lips. The active speech level was controlled during recording with a meter conforming to Recommendation P.56 (ITU-T, 1993). At the beginning of each logatom set recording, 20 seconds 1000Hz calibration tone and 30 seconds rose noise are inserted at a level equal to the mean active speech level.

For each measure point (Mp) (the place where the measure position was situated) a list of 300 logatoms has been prepared. The same logatom lists were never played for any condition day after day. The logatom lists at the four listener locations were recorded on the digital tape recorder. These recordings were played back over headphones to the subject afterward. This way of subjective measurements realization provides the same listening conditions for both traditional and with choice methods. In each room for each position of listener (Mp) and for each signal-to-noise ratio (SNR) the logatom intelligibility was obtained by averaging out the group of listeners results.

The listening tests were done in the studio of Institute of Telecommunications, Teleinformatics and Acoustics of Wroclaw University of Technology (Fig.9). Background noise level didn't exceed 40 dBA. Subjective measurements of logatom intelligibility were taken in conditions of binaural listening using headphones for the optimum speech signal level of 80 dBA.

Prior to the proper measurements the listening team was subjected to a 6 - hour training (two 3 - h sessions). The measurement sessions duration did not exceed 3h (together with 10 min breaks after every 20 min of listening).

Fig. 9. Speech quality subjective measurement stand.

The subjective logatom intelligibility measurement results, obtained for different speech to noise ratio, are presented graphically in fig. 10. The curves, representing the relationship between logatom intelligibility and the signal-to-noise ratio in a room were approximated by a four-degree polynomial calculated on the basis of the least-squarees method. The obtained relations are presented in fig. 10. which also includes values of correlation coefficient R^2 – a measure of the conformity between the polynomial and the results obtained from the subjective tests. As one can see there is very good agreement between the theoretical curve and the empirical results; the value of correlation coefficient R^2 exceeds 0.99 in each case.

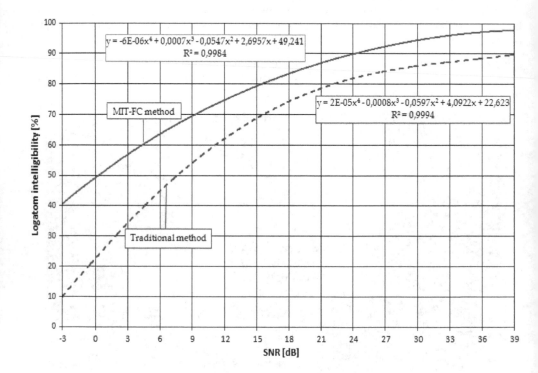

Fig. 10. Relationship between logatom intelligibility and signal-to-noise ratio (SNR) for rooms (Logatom intelligibility measured with traditional and MIT-FC method).

For a few randomly selected measuring points the distribution of $W_{n,k}$ values was compared with the normal distribution. The agreement between the $W_{n,k}$ distribution and the normal distribution were tested by applying the Kolmogorov-Smirnov test. It has been found that at significance level $\alpha = 0.05$ there are no grounds to reject the hypothesis about the goodness of fit of the distributions. Thus, it is reasonable to use average logatom hypothesis W_L as an estimator of the logatom intelligibility for a given measuring point.

The main goal of presented research was assessment if there exists a relation between measurement results of traditional method and modified intelligibility test with forced choice (MIT-FC) in rooms, and if such a relation exists, its finding out. The obtained results revealed that there exists monotonic relation between results of traditional method and MIT-FC. The results were presented on the surface: logatom intelligibility MIT-FC – traditional, and approximated by a fourth order polynomial. The obtained relations are presented in fig. 11. which also includes values of correlation coefficient R^2 – a measure of the conformity between the polynomial and the results obtained from the subjective tests.

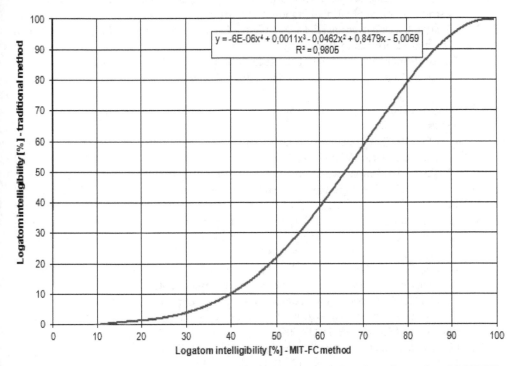

Fig. 11. Relationship between logatom intelligibility measured with traditional and MIT-FC method for rooms.

6. Conclusion

The presented MIT-FC method offers a simple, easy to use, stable, and fully automatized speech system to assessment of speech quality in rooms. The results of the experiments have shown that the MIT-FC method is very useful in the evaluation of speech quality in rooms. The time needed to carry out the measurement with MIT-FC method is the same as in traditional one but we obtain the results right after finishing the measurement process.

The experiments carried out in finding the relations between logatom intelligibility measured with traditional and semi-automatic with forced choice methods for the rooms have shown that there exist the multi-value and repetitive relation between them. It allows using both methods interchangeably and converting results between them.

The obtained relations are applied in the Institute of Telecommunications, Teleinformatics and Acoustics of Wrocław University of Technology to the design of subjective tests for the verification of results yielded by a new objective method based on automatic speech recognition techniques.

7. References

ANSI S 3.2, (2009), *Methods for Measuring the Intelligibility of Speech over Communication Systems*, American National Standards Institute.

ANSI S 3.5, (1997), *Methods for the Calculation of the Speech Intelligibility index (SII)*. American National Standards Institute

Basciuk K., Brachmanski S., (1997), The Automation of the Subjective Measurements of Logatom Intelligibility, Prep. 4407, *102-nd Convention AES*, Munich, Germany, March 22-25, 1997.

Brachmanski S., (1982), *Modulation Transfer Function (MTF) as a Measure of Polish Speech Transmission Quality* (in Polish), Ph. D. Thesis, Wrocław.

Brachmanski S., (1995), Choosing Optimum Number of Test Items in Subjective Logatom Measurements (in Polish). *Proc. XLII Open Seminar on Acoustics*, pp. 423-428, ISBN 83-902146-3-6, Warszawa-Bialowieza, September 12-15, 1995

Brachmanski S., (2002) The Automation of Subjective Measurements of Speech Intelligibility in Rooms, *112th Conv. AES*, Munich, Germany, May 10-13, 2002.

Brachmanski S., (2004) Estimation of Logatom Intelligibility with the STI Method for Polish Speech Transmitted via Communication Channels, *Archives of Acoustics*, Vol. 29, No. 4, (2004), pp. 555-562, , ISSN 0137-5075.

Brachmanski S., (2004), The Subjective Measurements of Speech Quality in Rooms, *Proc. of Subjective and Objective Assessment of Sound*, Poznan, September 1-3, 2004.

Brachmanski S., (2006), Experimental Comparison between Speech Transmission Index (STI) and Mean Opinion Scores (MOS) in Rooms,. *Archives of Acoustics*, Vol .31, No.4, (2006), pp. 171-176, ISSN 0137-5075.

Davies D.D., Davies C., (1989), Application of Speech Intelligibility to Sound Reinforcement. J. *Audio Eng. Soc.*, Vol. 37, No.12, (December 1989), pp. 1002-1018.

Farina A., (2001), Acoustic Quality of Theatres: Correlations between Experimental Measures and Subjective Evaluations, *Applied Acoustic*, Vol. 62, No 8, (August 2001), pp. 889-916, ISSN: 0003682X.

French, N.R., Steinberg, J.C., (1977), Factors Governing the Intelligibility of Speech Sounds, , In: *Speech Intelligibility and Speaker Recognition*, Hawley M. E., pp. 128 – 152, Dowden, Hutchinson & Ross Inc., ISBN 0-470-99303-0, Stroudsburg, Pennsylvania.

Houtgast T., Steeneken H.J.M., (1973), The Modulation Transfer Function in Room Acoustics as a Predictor of Speech Intelligibility, *J. Acoust. Soc. Am.* Vol. 54, No. 2, (August 1973), pp. 557-557, pp. 66-73, ISSN 0001-4966

Howard D.M, Angus J.A.S, *Acoustics and Psychoacoustics*, Elsevier-Focal Press, 2009, ISBN 978-0-240-52175-6, Oxford, UK.

IEC 60268-16, (2003), *Sound system equipment – Part16: Objective Rating of Speech Intelligibility by Speech Transmission Index*, International Electrotechnical Commission, Geneva, Switzerland.

ISO/TR 4870, (1991), *Acoustics – The Construction and Calibration of Speech Intelligibility Tests*. International Organization for Standardization.

ITU-T Recommendation P.56, (1993), *Objective Measurement of Active Speech Level*, International Telecommunication Union, Geneva, Switzerland.

ITU-T Recommendation P.800, (1996), *Method for Subjective Determination of Transmission Quality*, International Telecommunication Union, Geneva, Switzerland.

ITU-T Recommendation P.830, (1996), *Method for Objective and Subjective Assessment of Quality*, International Telecommunication Union, Geneva, Switzerland.

ITU-T Recommendation P.862, (2003), *Perceptual Evaluation of Speech Quality (PESQ), an Objective Method for End-To-End Speech Quality Assessment of Narrowband Telephone Networks and Speech Codecs*, International Telecommunication Union, Geneva, Switzerland.

ITU-T Recommendation P.863, (2011), *Methods for Objective and Subjective Assessment of Speech Quality. Perceptual Objective Listening Quality Assessment*, International Telecommunication Union, Geneva, Switzerland.

ITU-R Recom. BS.1116-1, (1997), *Method for the Subjective Assessment of Small Impairments in Audio Systems Including Multichannel Sound Systems*, International Telecommunication Union, Geneva, Switzerland.

Kryter, K.D., (1962), Methods for the Calculation and Use of the Articulation Index, *J. Acoust. Soc. Am.*, Vol. 34, No. 11, (November 1962), pp. 1689-1697, ISSN 0001-4966.

Lam P., Hongisto V., (2006), Experimental Comparison between Speech Transmission Index, Rapid Speech Transmission Index, and Speech Intelligibility Index, *J. Acoust. Soc. Am.*, Vol. 119, No. 2, (February 2006), pp.1106-1117, ISSN 0001-4966.

Majewski W., Basztura Cz., Myślecki W., (1988), Relation between Speech Intelligibility and Subjective Scale ff Speech Transmission Quality., *Proc. 7th FASE Symposium, Proceedings SPEECH'88*, 719-726, Edinburgh, August 22-26, 1988.

Majewski W., Myślecki W., Baściuk K., Brachmański S., (1998), Application of Modified Logatom Intelligibility Test in Telecommunications, Audiometry and Room Acoustics, *Proceedings 9th Mediterranean Electrotechnical Conference Melecon'98*, pp. 25-28, Tel-Aviv, Israel, May 18-20, 1998.

Majewski W., Myślecki W., Brachmański S., (2000), Methods of Assessing the Quality of Speech Transmission (in Polish), *Proceedings of 47th Open Seminar on Acoustics*, pp. 66-75, ISBN 83-914391-0-0, Rzeszow-Jawor, September 19-22, 2000.

Möller S. *Assessment and Prediction of Speech Quality in Telecommunications*, Kluwer Academic Publisher, ISBN 978-1-4419-4989-9, Dordrecht, Netherlands.

Polish Standard PN-90/T-05100, (1991), *Analog Telephone Chains. Requirements and Methods of Measuring Logatom Articulation* (in Polish), Polski Komitet Normalizacyjny (Polish Committee for Standardization), 1990, Warszawa, Poland

Polish Standard PN-V-90001, (1999), *Digital Communication Systems. Requirements and Methods for Measurement of Logatom Articulation* (in Polish), Polski Komitet Normalizacyjny (Polish Committee for Standardization), 1999, Warszawa, Poland

Sotschek J., (1976), Methoden zur Messung der Sprachgüte I: Verfahren zur Bestimmung der Satz- und der Wortverständlichkeit, *Der Fernmelde Ingenieur*, Vol. 30, No. 10, pp. 1-31.

Srinivasan S.H., (2004), Speech Quality Measure Based on Auditory Scene Analysis, *Proc. IEEE 6th Workshop on Multimedia Signal Processing*, pp. 371 – 374, ISBN: 0-7803-8578-0 September 29- October 1, 2004, Siena, Italy.

Steeneken H.J.M., Houtgast T., (1980), A Physical Method for Measuring Speech-Transmission Quality, *J. Acoust. Soc. Am.*, Vol. 67, No. 1, (January 1980), pp. 318 – 326, ISSN 0001-4966.

Steeneken H.J.M., Houtgast T., (2002), Validation of Revised STI Method, *Speech Communication*, Vol. 38, No. 3-4, (November 2002), pp.413-425, ISSN:0167-6393.

4

Automatic Restoration of Power Supply in Distribution Systems by Computer-Aided Technologies

Daniel Bernardon[1], Mauricio Sperandio[2],
Vinícius Garcia[1], Luciano Pfitscher[1] and Wagner Reck[2]
[1]Federal University of Santa Maria
[2]Federal University of Pampa
Brazil

1. Introduction

The need to improve quality and reliability of Power Systems has contributed to the advancement of research on Smart Grids and related topics. Some challenges that motivate the deployment of such researches include the increasing in energy consumption, environmental aspects, the integration of distributed and renewable generation, and new advances in computer aided technologies and automation. Smart Grids are characterized by a series of integrated technologies, methodologies and procedures for planning and operation of electrical networks. A survey of the main projects and researches related to Smart Grid is presented in (Brown, 2008).

Some of the main desired features in a Smart Grid are the low operating and maintenance costs and the ability to self-healing. In this context, utilities have concentrated significant efforts in order to improve the continuity of the supplied electrical energy, especially because of regulatory policies, besides the customer satisfaction and the improvement of the amount of energy available to commercial and industrial activities. However, supply interruptions are inevitable due to implementation of the expansion of the system, preventive maintenance on network equipment, or even by the action of protective devices due to defects.

A distribution network can have its topology changed by opening or closing switches, allowing to isolate faults and to restore the supply in contingency situations, and also in case of scheduled shutdowns. In addition, the change of topology allows a better load balancing between feeders, transferring loads of heavily loaded feeders to other feeders, thus improving the voltage levels and reducing losses and increasing levels of reliability.

The reconfiguration of a distribution network is considered an optimization problem in which the objective is to seek for one configuration, among several possible solutions, that leads to better performance, considering the ultimate goal of the reconfiguration and observing the network constraints. One factor that increases the complexity of the problem is the large number of existing switches in a real distribution network, which leads to a lot of different possible configurations to be analyzed.

The stages of planning the reconfiguration of distribution networks usually involve the collection of information of the network and load, the application of methodologies for estimating loads and calculation of power flows, and the application of the methodology for network optimization.

The behavior of the load is essential in the study of reconfiguration of distribution networks, because load variations cause demand peaks at different periods, and thus the reconfiguration at a particular period may not meet the objectives and constraints at later periods.

Several researches are related to the reconfiguration of distribution networks, based on mathematical methods, heuristics and artificial intelligence techniques, as shown in (Takur & Jaswanti, 2006). Generally, the reconfiguration aims to meet the maximal number of consumers as much as possible, or the maximum energy demand, with loss minimization and load balancing. The reduction of losses is often taken as the primary objective of the reconfiguration. When more than one objective function is defined, it is necessary to apply multicriteria methods for decision making. Among the restrictions, there is the need to maintain a radial network, and the operation within the limits of voltage and current capacity of equipment and conductors of the line. The coordination of the protection system of the new network configuration should be taken into consideration.

In recent years, new methodologies of reconfiguration of distribution networks have been presented, exploring the greater capacity and speed of computer systems, the increased availability of information and the advancement of automation, in particular, the SCADA (Supervisory Control and Data Acquisition). With the increased use of SCADA and distribution automation - through the use of switches and remote controlled equipment - the reconfiguration of distribution networks become more viable as a tool for planning and control in real time.

Act as soon as possible in a contingency situation may result in a minimum cost to the utility and consumers. When a fault is identified at any point of the network, the following procedures must be performed: identify the location of the problem; isolate the minimum portion of the distribution system by opening normally closed switches; restore the power supply to consumers outside the isolated block reclosing the feeder breaker and/or normally open switches; correct the problem; re-operate the switches to get back to the normal network status. To do that, generally a maintenance crew has to travel long distances, and may be affected by traffic jams and unfavorable accesses.

The automation of distribution systems, with the installation of remote controlled switches, plays an important role on reducing the time to implement a service restoration plan (Sperandio et al., 2007). These devices have shown to be economically viable due to the growth of a large number of automation equipment suppliers and new communication technologies.

The commitment of an efficient system to operate these devices is quite important for the utilities, aiming to guarantee the technical feasibility of the network reconfiguration with minimal time necessary to restore the energy supply of the affected consumers.

In this chapter, a methodology for the automatic restoration of power supply in distribution systems by means of remote controlled switches is presented. It includes a validation of the technical feasibility for the reconfiguration of the network in real time using computer simulations. Since there may be many configuration options with different gains, an algorithm based on a multiple criteria decision making method, the Analytic Hierarchical Process – AHP (Saaty, 1980), is employed to choose the best option for load transfers after contingencies. The AHP method has proven to be effective in solving multi-criteria problems, involving many kinds of concerns including planning, setting priorities, selecting the best choice among a number of alternatives and allocating resources. It was developed to assist in making decisions where competing or conflicting evaluation criteria difficult to make a judgment (Saaty, 1994). Additionally, the algorithms for load modeling and load flow are also presented, since they are essential for analysis of the maneuvers.

As a result, load transfers are carried out automatically, being preceded by computer simulations that indicate the switches to be operated and that ensure the technical feasibility of the maneuver, with the characteristics of agility and safety for the power restoration and in agreement with the Smart Grids concepts. The developed tool has been applied in a pilot area of a power utility in Brazil. The reduction in the displacement of maintenance teams and improving indices of continuity characterize the greatest benefits for the company, making a difference in the market and, consequently, generating economic and productivity gains.

2. Modeling of power load profiles

The most common information on power loads profile comes from utilities charges, based on measurements of monthly energy consumption. Unfortunately, these data are insufficient for the analysis of the distribution systems, since they do not reflect the daily power behavior. Therefore, a methodology for power load modeling is need, which usually employs typical load curves for their representation. A technique for building the typical load curves that has advantages in relation to the traditional statistical methods was presented in (Bernardon et al., 2008). It reduces the influence of random values and also the amount of measurements required to form a representative sample of the load types. Instead of using the simple average to determine the active and reactive power values for a typical load curve ordinate, the following equation is used:

$$X_t = \frac{1}{5} \cdot \left[2M\{X_t\} + 2Me\{X_t\} + Mo\{X_t\} \right] \qquad (1)$$

where:
X_t – active (P_t) or reactive (Q_t) power value for the time t of the typical load curve;
$M\{X_t\}$ – sample average;
$Me\{X_t\}$ – sample median (central number of a sample);
$Mo\{X_t\}$ – sample mode (most frequently occurring value of a sample).

According to the monthly power consumption data and economic activity developed, each customer is associated to a typical load curve. Based on the load factor (LF) and monthly energy consumption values (W), the maximum power demand (P_{max}) for a k group of consumers is calculated:

$$P_{max_k} = \frac{W_k}{T \cdot LF_k} \tag{2}$$

The typical curves utilized are normalized in relation to the maximum active demand, and the load curve for a k group of consumers is built by multiplying each ordinate by this value:

$$P_{kUt} = P_{max_k} \cdot P^*_{kUt}$$

$$P_{kSt} = P_{max_k} \cdot P^*_{kSt} \tag{3}$$

$$P_{kDt} = P_{max_k} \cdot P^*_{kDt}$$

$$Q_{kUt} = P_{max_k} \cdot Q^*_{kUt}$$

$$Q_{kSt} = P_{max_k} \cdot Q^*_{kSt} \tag{4}$$

$$Q_{kDt} = P_{max_k} \cdot Q^*_{kDt}$$

where:
P^*_{kt} and Q^*_{kt} - active and reactive power values, normalized for ordinate t of typical curve k;
U, S and D – Working days, Saturdays, Sundays/holidays, respectively.

The integral load curves for Working days, Saturdays and Sundays for the distribution transformer j is done through the sum of the load curves of the different groups of N_k consumers connected to it:

$$P_{jUt} = \sum_{i=1}^{N_k} P_{iUt} \; ; \; P_{jSt} = \sum_{i=1}^{N_k} P_{iSt} \text{ and } P_{jDt} = \sum_{i=1}^{N_k} P_{iDt} \tag{5}$$

$$Q_{jUt} = \sum_{i=1}^{N_k} Q_{iUt} \; ; \; Q_{jSt} = \sum_{i=1}^{N_k} Q_{iSt} \text{ and } Q_{jDt} = \sum_{i=1}^{N_k} Q_{iDt} \tag{6}$$

Thus, it is possible to consider the load levels corresponding to the period in which the failure occurred in the distribution network. Usually, the load transfers are analyzed considering the fault's time of the occurrence and the next four consecutive hours, thus ensuring the technical feasibility of the load transfers until the network returns to its original configuration.

3. Load flow method

A version of the classical backward/forward sweep method by (Kersting & Mendive, 1976) was performed to calculate the load flow in radial distribution networks. Since the electrical loads are defined by a constant power according to the applied voltage, the circulating current varies with the voltage drops. So, the solution is found only iteratively. The resulting procedure is described as follows:

Step 1. It is considered that the voltage in all points of the feeder is the same as the voltage measured in the substation bar. This information can be automatically received by the remote measurement systems installed at the substations. Voltage drops in the branches are not taken into account in this step.

Step 2. Active and reactive components of the primary currents absorbed and/or injected in the system by the electrical elements are calculated.

Step 3. The procedure to obtain the current in all network branches consists of two stages: (a) a search in the node set is performed adding the current values in the set of branches and (b) currents from the final sections up to the substation are accumulated.

Step 4. Voltage drops in primary conductors are determined.

Step 5. From the substation bus it is possible to obtain the voltage drops accumulated at any other part of the primary network, and, consequently, the voltage values at any point.

Step 6. The difference between the new voltage values for all nodes and the previous values is checked. If this difference is small enough comparing to a previously defined threshold, the solution for the load flow calculation was found and the system is said to be convergent. Otherwise, steps 2 to 6 are repeated, using the calculated voltages to obtain the current values. The threshold of 1% was chosen, because it leads to accurate values for the status variables without requiring too much processing time.

At the end of the process, the active and reactive powers and the technical losses in the primary conductors are defined for all branches of the feeder.

This load flow method was implemented in the proposed methodology for the automatic power supply restoration. It was used for analyzing the technical feasibility of the load transfers and the results were considered as constraints in the optimization procedure. That is, such transfers may neither cause an overload on the electrical elements (conductors and transformers), nor reach the pickup threshold of the protective devices, nor exceed the limits of voltage range of the primary network. The checking of the constraints is performed by considering the load profile compatible with the period of the failure.

4. Methodology for automatic operation of remote controlled switches to restore power supply

The logic for power restoration is presented considering the hypothetical example of the simplified distribution network illustrated in Figure 1. Normally close (NC) switch and normally open (NO) switches in Figure 1 are remotely controlled. Assuming that an outage has occurred in feeder FD-1, the procedure for electric power restoration is:

-    Fault downstream of NC-1 switch: in the event of this fault, the current values of short-circuit will be flagged online in the SCADA (Supervisory Control and Data Acquisition) system. So, it is assumed that the failure occurred downstream of NC-1 switch; then, this switch is operated automatically to isolate the defect.

- Fault upstream of NC-1 switch: in the event of this fault, the current values of short-circuit will not be flagged in the SCADA system. So, it is assumed that the failure occurred upstream of NC-1 switch, automatically operating the remote controlled switches to open NC-1 and to close NO-1 or NO-2 in order to transfer consumers downstream of NC-1 to another feeder.

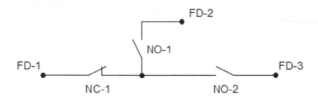

Fig. 1. Example of switches in a distribution network (FD – feeder, NC – normally closed switch, NO – normally open switch).

The technical and operational feasibility of load transfers using the remote controlled switches is verified by computer simulations. If there is more than one option of load transfer (e.g. to FD-2 or FD-3), the best option will be chosen considering the defined objective functions and constraints by a multiple criteria decision making algorithm. After this analysis, the developed tool automatically sends the necessary commands to maneuver the equipment.

Moreover, the automatic operation of remote controlled switches are carried out only after all attempts to restart the protection devices have been tried, i.e., they are done only in case of permanent fault, after a maximum of 3 minutes needed to complete the computer simulations and maneuvers of the switches since the instant of the fault identification. Figure 2 illustrates the architecture of the proposed system.

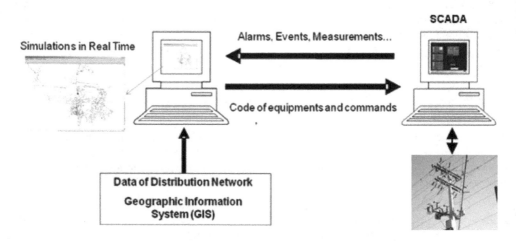

Fig. 2. Architecture of the developed system.

The flowchart of the proposed methodology is shown in Figure 3.

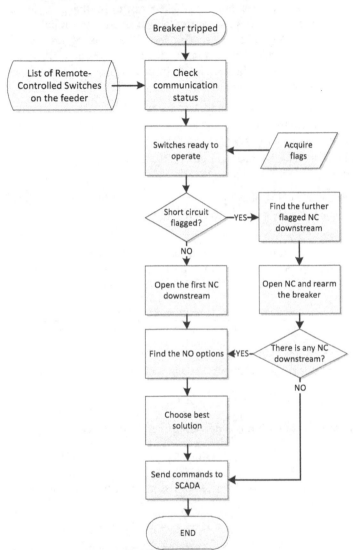

Fig. 3. Flowchart of the proposed methodology.

5. Methodology to determine the optimized reconfiguration based on the AHP method

After a contingency, the challenge is to decide which is the best load transfer scheme using the remote controlled switches among all possibilities, depending on previously defined objective functions and constraints. This is a multiple criteria decision making problem, since various types of objective functions can be considered.

The most common objectives are the maximization of restored consumers and of restored energy; however, generally it is not possible to optimize the grid for both objectives simultaneously. Furthermore, it is also important to ensure the reliability of distribution systems, through continuity indicators. The basic parameters are the SAIDI (System Average Interruption Duration Index) and the SAIFI (System Average Interruption Frequency Index), according to (Brown, 2009). In this approach, we adopt expected values based on the system's failure probability.

The constraints considered are the maximum loading of electrical elements, the protection settings and the allowable voltage drop in the primary network. Typically, the last two restrictions are the hard ones. On the other hand, a percentage of overloading of the network elements is acceptable in a temporary situation, assuming that the fault can be fixed in a couple of hours.

In our approach the following objective functions and constraints were defined to be used in the analysis of load transfers in case of contingencies:

Objective functions:
- Maximization of the number of restored consumers;
- Maximization of the amount of restored energy;
- Minimization of the expected SAIFI:

$$ESAIFI = \frac{\text{Expected Total Number of Customer Interruptions}}{\text{Total Number of Customers Served}} \tag{7}$$

Constraints:
- Current magnitude of each element must lie within its permissible limits:

$$|I_i| \le I_{i\,max} \tag{8}$$

- Current magnitude of each protection equipment must lie within its permissible limits:

$$|I_i| \le I_{jprot} \tag{9}$$

- Voltage magnitude of each node must lie within its permissible ranges:

$$V_{j\,min} \le V_j \le V_{j\,max} \tag{10}$$

where:
ESAIFI - expected value of system average interruption frequency (failures/year);
I_i - current at branch i;
$I_{i\,max}$ - maximum current accepted through branch i;
I_{jprot} - pickup current threshold of the protection device j;
V_j - voltage magnitude at node j;
$V_{j\,min}$ - minimum voltage magnitude accepted at node j;
$V_{j\,max}$ - maximum voltage magnitude accepted at node j.

The verification of the objective functions and constraints is made by calculation of the load flow for the various alternatives in real time. The ESAIFI is obtained by applying the classical equations of reliability during the process of calculating the load flow (Tsai, 1993).

Identifying the best option for load transfers is not simple since three objective functions are employed. For example, one particular option may have the largest number of consumers to be transferred, the other the largest amount of energy to be transferred, and the other the least expected value of consumers interrupted per year.

To solve this, the Analytic Hierarchical Process - AHP method was chosen, because of its efficiency in handling quantitative and qualitative criteria for the problem resolution. The first step of the AHP is to clearly state the goal and recognize the alternatives that could lead to it. Since there are often many criteria considered important in making a decision, the next step in AHP is to develop a hierarchy of the criteria with the more general criteria at the top of it. Each top level criteria is then examined to check if it can be decomposed into subcriteria.

The next step is to determine the relative importance of each criterion against all the other criteria it is associated with, i.e., establish weights for each criterion. The final step is to compare each alternative against all others on each criterion on the bottom of the hierarchy. The result will be a ranking of the alternatives complying with the staged goal according to the defined hierarchy of the criteria and their weights (Baricevic et al., 2009).

In the proposed approach, the main criterion is to choose the best option for load transfers and the subcriteria are the proposed objective functions. The alternatives are the options for load transfers.

An example of the AHP algorithm is defined in (Saaty, 1980):

1. The setup of the hierarchy model.
2. Construction of a judgment matrix. The value of elements in the judgment matrix reflects the user's knowledge about the relative importance between every pair of factors. As shown in Table 1, the AHP creates an intensity scale of importance to transform these linguistic terms into numerical intensity values.

Intensity of Importance	Definition
1	Equal importance
3	Weak importance of one over another
5	Essential or strong importance
7	Very strong or demonstrated importance
9	Absolute importance
2, 4, 6, 8	Intermediate values between adjacent scale values

Table 1. Intensity scale of importance (Yang & Chen, 1989).

Assuming $C_1, C_2, ..., C_n$ to be the set of objective functions, the quantified judgments on pairs of objectives are then represented by an n-by-n matrix:

$$M = \begin{matrix} C_1 \\ C_2 \\ \vdots \\ C_n \end{matrix} \begin{bmatrix} 1 & a_{12} & \cdots & a_{1n} \\ 1/a_{12} & 1 & \cdots & a_{2n} \\ \vdots & \vdots & \ddots & \vdots \\ 1/a_{1n} & 1/a_{2n} & \cdots & 1 \end{bmatrix} \qquad (11)$$
$$ C_1 \qquad C_2 \quad \cdots \quad C_n$$

Where n is the number of objective functions and the entries $a_{i,j}$ ($i, j = 1, 2, ..., n$) are defined by the following rules:

- if $a_{i,j} = \alpha$, then $a_{j,i} = 1/\alpha$, where α is an intensity value determined by the operators, as shown in Table 1;
- if C_i is judged to be of equal relative importance as C_j, then $a_{i,j} = 1$, and $a_{j,i} = 1$; in particular, $a_{i,i} = 1$ for all i.

3. Calculate the maximal eigenvalue and the corresponding eigenvector of the judgment matrix M. The weighting vector containing weight values for all objectives is then determined by normalizing this eigenvector. The form of the weighting vector is as follows:

$$W = \begin{bmatrix} w_1 \\ w_2 \\ \vdots \\ w_n \end{bmatrix} \qquad (12)$$

4. Perform a hierarchy ranking and consistency checking of the results. To check the effectiveness of the corresponding judgment matrix an index of consistency ratio (CR) is calculated as follow (Saaty & Tran, 2007):

$$CR = \frac{\left(\dfrac{\lambda_{max} - n}{n - 1} \right)}{RI} \qquad (13)$$

where:
λ_{max} = the largest eigenvalue of matrix M;
RI = random index.

A table with the order of the matrix and the RI value can be found in Saaty, 1980. In general, a consistency ratio of 0.10 or less is considered acceptable.

The AHP method was implemented in the proposed methodology. Its application is presented considering the example of distribution network illustrated in Figure 4. Normally close (NC) switch and normally open (NO) switches are remotely controlled.

The main goal is to define the best option for load transfer, considering that an outage has occurred in the feeder F1 (fault upstream of NC-1). In this case, there are two transfer options:

a. open the NC-1 switch and close the NO-1 switch; or
b. open the NC-1 switch and close the NO-2 switch.

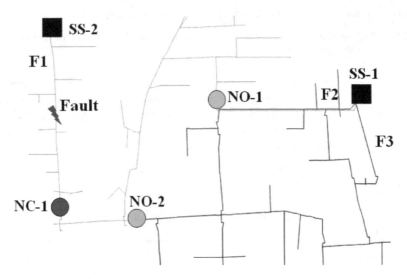

Fig. 4. Distribution network with three feeders.

First, the judgment matrix was obtained:

$$M = \begin{array}{c} \\ C_1 \\ C_2 \\ C_3 \end{array} \begin{array}{ccc} C_1 & C_2 & C_3 \\ \begin{bmatrix} 1 & 3 & 5 \\ 1/3 & 1 & 3 \\ 1/5 & 1/3 & 1 \end{bmatrix} \end{array} \tag{14}$$

where:
M = judgment matrix;
C_1 = number of restored consumers;
C_2 = amount of restored energy;
C_3 = expected number of interrupted consumers per year.

Thus, the weight values for the three objective functions were determined:

$$W = \begin{bmatrix} 0.64 \\ 0.26 \\ 0.10 \end{bmatrix} \tag{15}$$

where:
λ_{max} = 3.07.

The consistency ratio (CR) was calculated by Equation 13:

$$CR = \frac{\left(\dfrac{3.07 - 3}{3 - 1} \right)}{0.52} = 0.0673 \tag{16}$$

The consistency ratio is lower than 0.10 and it is considered acceptable.

Tables 2 and 3 show the results of the analysis for each load transfer. In this example, there is no violation on the constraints.

Options	Number of Restored Consumers	Amount of Restored Energy (kW)	Expected Number of Interrupted Consumers per Year
1 (open NC-1 and close NO-1)	14,000	1,930.00	1,800
2 (open NC-1 and close NO-2)	14,000	1,930.00	2,300
Base Selected	14,000	1,930.00	1,800

Table 2. Results of the analysis for each load transfer.

Options	Number of Restored Consumers	Amount of Restored Energy	Expected Number of Interrupted Consumers per Year
1	1.00	1.00	1.00
2	1.00	1.00	0.78

Table 3. Normalized values of Table 2.

The results using AHP method were obtained by Equation 17:

$$\begin{bmatrix} Op.1 \\ Op.2 \end{bmatrix} = \begin{bmatrix} 1.00 & 1.00 & 1.00 \\ 1.00 & 1.00 & 0.78 \end{bmatrix} \cdot \begin{bmatrix} 0.64 \\ 0.26 \\ 0.10 \end{bmatrix} = \begin{bmatrix} 1.00 \\ 0.98 \end{bmatrix} \tag{17}$$

According to the proposed method the option "1" is considered the best solution. Thus, the system performs the commands to make the load transfer, open NC-1 and close NO-1, without violating the set of constraints.

6. Experimental analysis

To verify the performance of the proposed methodology several case studies were carried out in the concession area of a power utility in Brazil. The developed methodology was applied on the distribution system shown in Figure 5, which has 20 distribution substations, 125 feeders, 214 remote controlled switches, and 523,619 consumers.

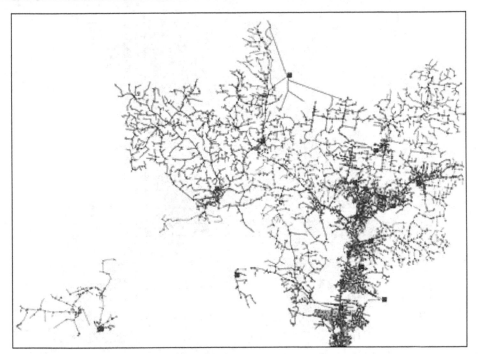

Fig. 5. Distribution network of a power utility in Brazil.

Figures 6 and 7 show the results for the calculation of load flow and the typical load curves used, respectively:

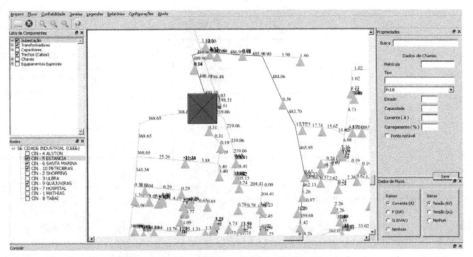

Fig. 6. Results for the calculation of load flow with the indication of values of current in branches and voltage in nodes.

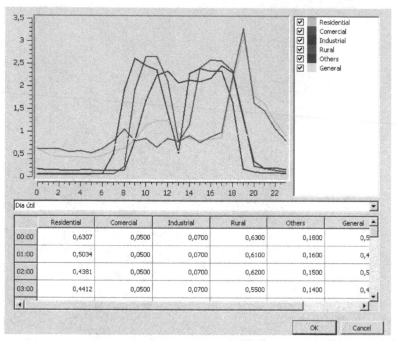

Fig. 7. Typical load curves used.

Table 4 shows the results obtained by the application of this methodology in case of outage of a feeder when considering the power restoration time.

Description	Mean Time to Restore Energy			
	Faults upstream the NC switch		Faults downstream the NC switch	
	Clients upstream the switch	Clients downstream the switch	Clients upstream the switch	Clients downstream the switch
Before installing the remote controlled switches	1h54min	58min	43min	1h34min
After installing the remote controlled switches	1h54min	0min	0min	1h34min
Reduction	-	58min	43min	-

Table 4. Results obtained with the use of remote controlled switches.

It should be noted that a reduction of approximately 30 % on the annual SAIDI index of this system is expected, assuming the number of faults in the main feeder.

Figures 8 and 9 show the picture of a pole top remote controlled switch been installed and the screen of the SCADA system, respectively:

Fig. 8. Installation of a remote controlled switch.

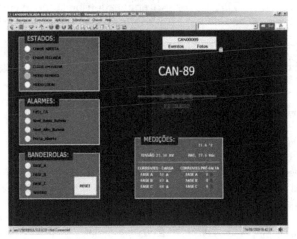

Fig. 9. Screen of the SCADA system.

7. Conclusion

This work presented a methodology developed for automatic power restoration system, which operates remote controlled switches in the distribution network. It was show how to assess the technical feasibility of the load transfers in real time by means of computer simulations, and how the best maneuver option to execute after a contingency is defined based on the AHP multicriterial method. The system will automatically handle the load transfers in accordance with the defined objective functions, without violating the established constraints. Case studies with real data from utilities were conducted to evaluate the performance of software developed presenting satisfactory results.

8. Acknowledgment

The authors would like to thank the technical and financial support of AES Sul Distribuidora Gaúcha de Energia SA, Conselho Nacional de Desenvolvimento Científico e Tecnológico (CNPq), and Coordenação de Aperfeiçoamento de Pessoal de Nível Superior (CAPES).

9. References

Baricevic, T., Mihalek, E., Tunjic, A. & Ugarkovic, K. (2009). AHP Method in Prioritizing Investments in Transition of MV Network to 20kV, *CIRED 2009 - 20th International Conference on Electricity Distribution*, pp. 1-4, Jun. 2009.

Bernardon, D.P., Comassetto, L. & Canha, L.N. (2008). Studies of parallelism in distribution networks served by different-source substations, *Electric Power Systems Research*, Elsevier, v. 78, p. 450-457, 2008.

Brown, R.E. (2008). Impact of Smart Grid on distribution system design, *IEEE Power and Energy Society General Meeting - Conversion and Delivery of Electrical Energy in the 21st Century*, pp.1-4, 2008.

Brown, R.E. (2009). *Electric Power Distribution Reliability*, CRC Press, Second Edition, ISBN 978-0-8493-7567-5, New York, 2009.

Kersting, W.H. & Mendive, D.L. (1976). An application of ladder network theory to the solution of three-phase radial load-flow problems, *IEEE Power Engineering Society Winter Meeting*, vol. A76 044-8, pp. 1-6, 1976.

Saaty, T.L. & Tran, L.T. (2007). On the invalidity of fuzzifying numerical judgments in the Analytic Hierarchy Process, *Mathematical and Computer Modelling*, vol. 46, pp. 962–975, 2007.

Saaty, T.L. (1980). *The Analytic Hierarchy Process: Planning, Priority Setting, Resource Allocation*, McGraw-Hill, ISBN 0-07-054371-2, New York, 1980.

Saaty, T.L. (1994). Highlights and Critical Points in the Theory and Application of the Analytic Hierarchy Process, *European Journal of Operational Research*, vol. 52, pp. 426-447, 1994.

Sperandio, M., Coelho, J., Carmargo, C.C.B., et al. (2007). Automation Planning of Loop Controlled Distribution Feeders, *2nd International Conference on Electrical Engineering (CEE'07)*, Coimbra, 2007.

Thakur, T. & Jaswanti (2006). Study and Characterization of Power Distribution Network Reconfiguration, *Proc. 2006 IEEE Transmission & Distribution Conference and Exposition: Latin America*, pp. 1-6.

Tsai, L. (1993). Network reconfiguration to enhance reliability of electric distribution systems, *Electric Power Systems Research*, Elsevier, no. 27, pp. 135-140, 1993.

Yang, H.T. & Chen, S.L. (1989). Incorporating a multi-criteria decision procedure into the combined dynamic programming/production simulation algorithm for generation expansion planning, *IEEE Transaction Power System*, vol. 4, pp. 165–175, Feb. 1989.

5

Automation in Aviation

Antonio Chialastri
Medicair, Rome,
Italy

1. Introduction

An aircraft landed safely is the result of a huge organizational effort required to cope with a complex system made up of humans, technology and the environment. The aviation safety record has improved dramatically over the years to reach an unprecedented low in terms of accidents per million take-offs, without ever achieving the "zero accident" target. The introduction of automation on board airplanes must be acknowledged as one of the driving forces behind the decline in the accident rate down to the current level.

Nevertheless, automation has solved old problems but ultimately caused new and different types of accidents. This stems from the way in which we view safety, systems, human contribution to accidents and, consequently, corrective actions. When it comes to aviation, technology is not an aim in itself, but should adapt to a pre-existing environment shared by a professional community.

The aim of this paper is to show why, when and how automation has been introduced, what problems arise from different ways of operating, and the possible countermeasures to limit faulty interaction between humans and machines.

This chapter is divided into four main parts:

1. Definition of automation, its advantages in ensuring safety in complex systems such as aviation;
2. Reasons for the introduction of onboard automation, with a quick glance at the history of accidents in aviation and the related safety paradigms;
3. Ergonomics: displays, tools, human-machine interaction emphasizing the cognitive demands in fast-paced and complex flight situations;
4. Illustration of some case studies linked to faulty human-machine interaction.

2. What is automation

According to a shared definition of automation, the latter may be defined in the following way: "Automation is the use of control systems and information technologies to reduce the need for human work in the production of goods and services". Another plausible definition, well-suited the aviation domain, could be: "The technique of controlling an apparatus, a process or a system by means of electronic and/or mechanical devices that replaces the human organism in the sensing, decision-making and deliberate output" (Webster, 1981).

The Oxford English Dictionary (1989) defines automation as:

1. Automatic control of the manufacture of a product through a number of successive stages;
2. The application of automatic control to any branch of industry or science;
3. By extension, the use of electronic or mechanical devices to replace human labour.

According to Parasumaran and Sheridan, "automation can be applied to four classes of functions:

1. Information acquisition;
2. Information analysis;
3. Decision and action selection;
4. Action implementation."

Information acquisition is related to the sensing and registration of input data. These operations are equivalent to the first human information processing stage, supporting human sensory processes. If we adopt a decision-making model based on perception, identification, mental process, decision, action, follow-up and feedback, information acquisition could be likened to the first step: perception. Let's imagine a video camera and the aid it offers in monitoring activity. It helps to replace continuous, boring, monotonous human observation with reliable, objective and detailed data on the environment.

Automation may handle these functions, as it is more efficient in detecting compared to humans, while – at the same time – it offers the possibility of positioning and orienting the sensory receptors, sensory processing, initial data pre-processing prior to full perception, and selective attention (e.g.: the focus function in a camera).

Information analysis is related to cognitive functions such as working memory and inferential processes. It involves conscious perception and manipulation of processed items. It allows for quick retrieval of information in the working memory. In aviation, this kind of system is broadly used to provide pilots with predictive information, such as how much fuel will be available at destination, where the top of climb or top of descent will be in order to optimize the flight path, and so forth.

With regard to decision and action selection, automation is useful because it involves varying levels of augmentation or replacement of human decision-making with machine decision-making. It is generally acknowledged that human decision-making processes are subject to several flaws, among them a tendency to avoid algorithmic thought, a biased development of pros and cons based on the laws of logic, a partial view of the overall system and, often, the heavy influence of emotions.

The fourth stage involves the implementation of a response or action consistent with the decision taken. Generally, it this stage automation replaces the human hands or voice. Certain features in the cockpit allow automation to act as a substitute for pilots. For instance, this occurs when – following an alert and warning for windshear conditions – the automation system detects an imminent danger from a power setting beyond a pre-set threshold. In this case, the autopilot automatically performs a go-around procedure, which avoids a further decline in the aircraft's performance.

Besides being applicable to these functions, automation has different levels corresponding to different uses and interactions with technology, enabling the operator to choose the optimum level to be implemented based on the operational context (Parasumaran, Sheridan, 2000). These levels are:

1. The computer offers no assistance; the human operator must perform all the tasks;
2. The computer suggests alternative ways of performing the task;
3. The computer selects one way to perform the task and
4. Executes that suggestion if the human operator approves, or
5. Allows the human operator a limited time to veto before automatic execution, or
6. Executes the suggestion automatically then necessarily informs the human operator, or
7. Executes the suggestion automatically then informs the human operator only if asked.
8. The computer selects the method, executes the task and ignores the human operator.

3. History of accidents

Automation in the aviation world plays a pivotal role nowadays. Its presence on board airplanes is pervasive and highly useful in improving the pilots' performance and enhancing safety. Nevertheless, certain issues have emerged in the recent past that evidence automation misuse by pilots. This could depend on a series of factors, among them human performance, capabilities and limitations on one side, and poor ergonomics on the other.

We should first investigate the reasons leading to the introduction of onboard automation.

During the Fifties and Sixties, the main causes of aviation accidents were believed to be related to the human factor. The immediate cause of an accident was often to be found in "active failures", e.g. loss of control of the aircraft in which pilots failed to keep the aircraft under control, reaching over-speed limits, stalling, excessive bank angles, etc.

In these cases the root cause was a flawed performance that eventually caused the loss of control (the effect). Factors related to human performance, e.g. the impact of fatigue, attention, high workload sustainability, stress mismanagement, etc. were consequently addressed. Technological solutions were sought to help pilots manage these factors. Innovation at that time eventually led to the introduction of the auto-pilot, auto-throttle, flight director, etc. After the mid-Fifties, as a result of these innovations, the accident curve dropped sharply.

Looking at the graph below, we can clearly notice the impact of such innovations on flight safety. The vertical axis corresponds to the number of accidents per million take-offs, while the horizontal axis corresponds to the relative decades.

As we can observe, after a dramatic improvement the accident curve rose again during the mid-Seventies. Aviation safety experts were faced with accidents involving a perfectly functioning aircraft, with no evidence whatsoever of malfunctions. In these cases (known as "Controlled Flight Into Terrain" - CFIT), the aircrafts were hitting obstacles with the pilots in full control. The accidents were caused by loss of situational awareness, either on the horizontal or on the vertical path. The evidence showed that an improper interaction between pilots was the main cause behind the accident, so this time the solution came from psychology. Human factor experts developed techniques and procedures to enhance cooperation between pilots, and specific non-technical training aspects became mandatory for pilots. For instance, a Cockpit Resource Management (CRM) course is nowadays mandatory in a pilot's curriculum: its topics may include leadership, cross-checking and criticizing fellow colleagues, assertiveness, resolution of conflicts, communication skills, etc.

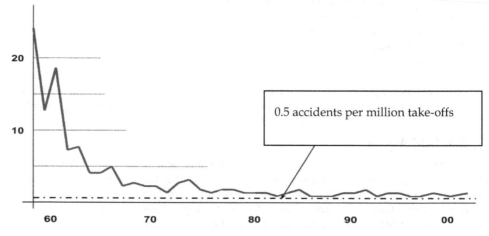

Source: ICAO doc. 9683/950. Accident rate over the years

In the last decade, the pendulum has swung back to loss of control as a major cause of aviation accidents, however, compared to the accidents occurring during the Fifties, the factors leading to loss of control appear to be different. Whereas in the beginnings of aviation, human performance was impaired by "under-redundancy", that is, insufficient aids available to pilots for avoiding the effects of factors like fatigue, distraction, workload and stress which reduced the pilots' performance, nowadays many domain experts are pointing at possible cases of "over-redundancy". This means that increasing automation might be putting the pilot out-of-the-loop, thus causing reduced situational awareness, automation complacency or over-confidence and loss of skills, due to lack of practice in manually flying the aircraft. As a result, pilots may not be able to regain control once automation has failed, or may be incapable of effectively monitoring the performance of automated systems (and questioning it when required).

The safety philosophy behind the adoption of increasing onboard automation is based on the assumption that human error is the main cause of accidents. Therefore, since the human (*liveware*) component of the system is the flawed link in the accident chain, we ought to look for a substitute capable of handling the tasks once performed by pilots. This is partially true, as we'll see later on. It is first necessary to understand what are the pros and cons of human contribution to safety, at what levels of operation does automation offer undoubted advantages and where the latter should end to leave room for pilots' decisions. Pilots and machines are not alternatives, but complementary factors in ensuring flight safety. Achieving the correct balance between these components of the aviation domain benefits safety, since the role of technology appears to have reached a standstill. Until the mid-1990s, it appears that pilots tried to adjust their behaviour to a given challenge (new automation), which was conceived regardless of their actual need and deeply rooted habits. In fact, one of the main drivers of the recent cockpit design philosophy was the reduction of costs related to better performances, lower fuel consumption, cheaper maintenance and flexible pilot training. Concern for the adaptability of pilots to these new solutions only came at a later stage and following some severe mishaps. To achieve this balance, we'll briefly analyse what levels of operation are involved in flight and where automation – primarily conceived to replace certain human operator tasks – should give way to the pilot's intervention.

4. Skill, rules and knowledge

According to a paradigm proposed by Rasmussen and also developed by James Reason, human activity can be grouped into three main fields: skill-based action, rule-based action and knowledge-based action.

The first field is based on the human capability to accomplish physical tasks, such as providing correct input to flight control (so-called "stick and rudder training"), responding to external stimuli in a quick and consistent way, and coordinating the body in order to obtain a desired result. It is mainly an area in which the psycho-physiological aspect is paramount. Moreover, we could also include monitoring tasks in this field, such as detection, identification and response to external signals stemming from habits (body automatism or conditional reaction). Automation has played an important role at this level by replacing human performance rather well. As a result, autopilots, auto-throttle (and later on, auto-thrust computers) have come to gradually replace pilots in "hand-flying". Generally speaking, an autopilot can tolerate workloads that are hardly sustainable by a pilot. Let's imagine an oceanic flight during the night; an autopilot is able to maintain (with no effort at all) altitude, speed, track and so forth, whereas pilots are subject to tiredness, attention lapses, distraction, etc. On the other hand, the systematic replacement of basic flying skills has led to the erosion of competence, because, as Germans put it: *"Die übung macht den meister"* ("Use makes master"). In the U.S. the FAA (Federal Aviation Authority) has suggested adopting "back to basics training", in which pilots are taught how to fly without the help of automatisms and how to retrieve the elementary notions of aerodynamics, in order to avoid grossly misreading altitude, speed and power.

The second level – the rule base action – is the conceptual layer. It indicates compliance with the rules, norms, laws, and everything laid down in the official documentation. It is unproblematic to apply a given rule whenever conditions warrant it. This is the case of a limit set for a device, e.g. the maximum temperature for operating an engine (EGT). When the upper limit is exceeded, something happens: red indications on instruments, alarms, flashing light on to attract the pilot's attention, automatic exclusion of the failed system and so forth. A machine can easily detect whether the operating conditions are normal or abnormal, by matching the real values with an operating envelope. Since the pilot may forget some rules, apply them incorrectly, or fail to apply valid norms, certain functions (especially those relating to the monitoring activity that induces boredom and complacency) are assigned to automation. It is a consequence of automation, therefore, that the flight engineer is no longer required in the cockpit. Some problems were initially detected in the normal flying activity of newly designed cockpits, since two pilots were required to manage a three-pilot cockpit, with automation playing the role of a "silent crew-member".

The third level – the knowledge-base – includes the sound judgement of pilots in deciding if and when the given rules are applicable. This implies the notion of a complex system. Complexity is evident at every level of reality, from physics to biology, from thermodynamics to meteorology (Morin, 2001). Different conceptions of complexity emerge in the current scientific debate, but generally speaking, we may highlight some commonalities between the different theories: refusal of reductionism, different level of system description (be it physical or biological or even man-made) according to the level of observation, emergent proprieties, etc.

Since aviation is a complex system made up of complex subsystems such as humans, technology and the environment, it is almost impossible to govern everything in advance through rules and norms. There will always be a mismatch between the required task and the final outcome (Hollnagel, 2006). The resilience engineering approach to safety is aware of this complexity and focuses on the ambivalent role – in such a system – of man, who simultaneously constitutes a threat and resource in coping with unexpected events, unforeseeable situations and flawed procedures. Much of this activity, which continuously and strategically adapts the means to the goal, is undetected either by the top management, or by the front-line operators themselves (pilots). These micro-corrections are so pervasive that the person involved in accomplishing a task fails to even realize how much he/she deviates from a given rule. The front-end operator should always seek to compromise between efficiency and thoroughness. Hollnagel calls this compromise the ETTO (Efficiency, Thoroughness Trade Off) principle. The paradox emerging from the blind application of rules – the so-called "white strike" or "working by the rule" – is that it leads to a paralysis of the entire production activity (Hollnagel, 2009). The effects of such shortcuts during normal operations are another area of concern affecting flight safety, due to the systems' opacity, the operator's superficial knowledge, uncertainties and ambiguities of the operational scenario. If we turn to the initial introduction of onboard automation, we may detect some commonalities regarding the way pilots have coped with the innovations.

5. Evolution of automation

Automation seems to follow an evolutionary path rather than a revolutionary approach. Its adoption on board aircrafts does not respond to the planned purpose of enhancing safety "from scratch" in a consistent way, but rather resembles a biological organism trying to continuously adapt to the challenges posed by its environment (fly-fix-fly).

This trial-and-response approach can be observed regardless of the fact that innovation introduced on board generally lags a step behind the overall level achieved by the industry. In fact, one of the requirements for a certain technology to be implementable in the aviation domain is its reliability; it is preferable to have a slower yet reliable system rather than a high-speed one that is not completely tested or tried in an operational environment.

We may identify three main generations of onboard automation systems: mechanical, electrical and electronic.

In the beginnings of commercial flight there were no instrumental aids to help pilots to fly. A piece of string was attached to the wing to indicate whether the airflow over it was sufficient to sustain flight. Later on, the first anemometers and altimeters were introduced to indicate pilots the airspeed and altitude, respectively. These were the first steps toward the "virtualization of the environment" (Ralli, 1993). The invention of the pneumatic gyroscope (replaced, shortly after, by the electric gyroscope), used to stabilize an artificial horizon, helped pilots to understand their situation even in meteorological conditions characterised by extremely poor visibility, while at the same time preventing dangerous vestibular illusion (a false sense of equilibrium stemming from the inner ear). These simple instruments were merely capable of providing basic indications. Early signs of automation were introduced on board aircrafts during the decade from 1920 to 1930, in the form of an autopilot based on a mechanical engineering concept that was designed to keep the aircraft flying straight: a very basic input to control the flight at a "skill" level. Moreover, as airplanes became bigger and bigger, it became necessary to apply some form of amplification of the pilots' physical force, because of the airflow over large aerodynamic surfaces. Servo-mechanisms were introduced on board, alongside certain devices aimed at facilitating perception of the force acting on such surfaces (artificial feel load, mach trim compensator) and absorbing the effects of the so-called Dutch-roll, an abnormal behaviour whereby the airplane yaws and oscillates in an uncoordinated manner (yaw damper). This (mechanical) innovation was the first of multiple steps that began widening the gap between the pilot's input (action on the yoke) and the final outcome (aerodynamic movement). Instead of direct control, with the yoke mechanically attached to the ailerons, airplanes began to be constructed with a series of mechanisms intervening between the pilot's input and the expected output. In this case, the virtualization of flight controls accompanied the parallel virtualization of flight instruments introduced by the artificial horizon (Attitude Display Indicator). At this stage, automation aided pilots mainly in their skill-based activity.

The second generation of automation included electric devices replacing the old mechanisms. Electric gyroscopes instead of pneumatic ones, new instruments such as the VOR (Very High Frequency Omni-directional Range) to follow a track based on ground aids, the ILS (Instrumental Landing System) to follow a horizontal and a vertical path till the runway threshold, and so forth. The 1960s saw plenty of innovations introduced on board aircrafts that enhanced safety: electric autopilots, auto-throttle (to manage the power setting in order to maintain a selected speed, or a vertical speed), flight directors (used to show pilots how to manoeuvre to achieve a pre-selected target such as speed, path-tracking and so forth), airborne weather radars, navigation instruments, inertial platforms, but also improved alarming and warning systems capable of detecting several parameters of engines and other equipment. Whereas the first generation of automation (mechanical) managed the pilot's skill-based level, the second generation managed the skill-based and rule-based levels previously assigned to pilots. The airplane systems were monitored through a growing number of parameters and this gave rise to a new concern: the inflation of information with hundreds of additional gauges and indicators inside the cockpit, reaching almost 600 pieces (Boy, 2011). At this stage, pilots used the technology in a tactical manner. In other words, their inputs to automation were immediately accessible, controllable and monitored in the space of a few seconds. For example, if the pilot wanted to follow a new heading, he would use a function provided by the autopilot. The desired heading value was selected in the glare-shield placed in front of the pilot's eye and the intended outcome would be visible within a few seconds: the airplane banked to the left or right to follow the new heading. End of the task. At this stage, automation helped pilots also at a rule-based level, since monitoring of thousands of parameters required efficient alarming and warning systems, as well as recovery tools.

The third generation of innovation involved electronics, and was mainly driven by the availability of cheap, accessible, reliable and usable technology that invaded the market, bringing the personal computer into almost every home. The electronic revolution occurring from the mid-80s also helped to shape the new generation of pilots, who were accustomed to dealing with the pervasive presence of technology since the early years of their life. Electronics significantly helped to diminish the clutter of instruments on board and allowed for replacing old indicators - gauges in the form of round-dial, black and white mechanical indicators for every monitored parameter - with integrated coloured displays (e.g: CRT: Cathode Ray Tube, LCD: Liquid Crystal Display) capable of providing a synthetic and analytic view of multiple parameters in a limited area of the cockpit.

It is worthy mentioning that the type of operations implied by the Flight Management System shifted from tactical to strategic. In fact, whilst in the previous electrical automation stage, pilots were accustomed to receiving immediate feedback visible shortly after the entered input, in the new version, a series of data entered by the operator would show their effects at a distance of hours. The data was no longer immediately accessible and visible, therefore this new way of operating placed greater emphasis on crew co-ordination, mutual cross-checking, operational discipline, not only in the flying tasks but also in the monitoring activity. The Flight Management System database contains an impressive amount of data, from navigational routes, to performance capabilities and plenty of useful information that can be retrieved from pilots. Further on we analyze the traps hidden behind this kind of automation.

However, it is important to point out that the actual discontinuity introduced by this generation of automation: was the notion of electronic echo-system. Compared to the past, when pilots were acquainted with the inner logic of the systems they used, their basic components, and normal or abnormal procedures for coping with operational events, in the new cockpit pilots are sometimes "out of the loop". This occurrence forces them to change their attitude towards the job. On an airplane such as the A-320 there are almost 190 computers located in almost every area of the fuselage. They interact with each other without the pilot being aware of this interaction. Every time the pilot enters an input to obtain a desired goal (e.g. activating the hydraulic system), he/she starts a sequence in which not only the selected system is activated, but also a number (unknown to the pilot) of interactions between systems depending on the flight phase, operational demands, the airplane's conditions, etc.

The unmanageable complexity of the "electronic echo-system" is a genuine epistemological barrier for the pilot. Whereas before, the pilot had thorough knowledge of the entire airplane and could strategically operate in a new and creative manner whenever circumstances required, the evolution of the cockpit design and architecture has brought about a new approach to flight management that is procedural and sequential. Only actions performed in accordance with computer logic and with the given sequence are accepted by the system. In acting as a programmer who cannot perform tasks that are not pre-planned by the computer configuration, the pilot has lost most of his expertise concerning the hardware part of the system (the aircraft). He too is constrained by the inner logic which dictates the timing of operations, even in high-tempo situations. Consequently, in the international debate on automation and the role of pilots, the latter are often referred to as "system operators". This holds true up to a certain point but, generally speaking, it is an incorrect assessment. Recalling what has been said about the levels of operation, we may say that at the skill-based level, pilot have become system operators because flying skills are now oriented to flight management system programming. As a flight instructor once said: "we now fly with our fingers, rather than with a hands-on method". What he meant was that the pilot now appears to "push-the-button" rather than govern the yoke and throttle.

At the skill-based level, automatisms may be in charge for the entire flight, relegating the pilot to a monitoring role. However, at a rule-based level, computers manage several tasks once accomplished by pilots, including monitoring the pressurization system, air conditioning system, pneumatic system and so forth. This statement is no longer valid when referred to the knowledge-based level. Here, the pilot cannot be replaced by any computer, no matter how sophisticated the latter is. Sound judgment of an expert pilot is the result of a series of experiences in which he/she fills the gap between procedures and reality. Since this paper discusses automation rather than the human factor in aviation, anyone wanting to investigate concepts such as flexibility, dealing with the unexpected, robustness, etc. may refer to authors adopting a Resilience Engineering approach (Hollnagel, 2008) (Woods, Dekker, 2010).

The fact that pilots are no longer so acquainted with the airplane as in the past, has led them to only adopt the procedural way to interact with the airplane. This is time-consuming, cognitively demanding, and above all, in some cases it may lead to miss the "big picture", or situation awareness. This new situation introduces two major consequences: automation intimidation (ICAO, 1998) and a restriction of the available tools for coping with unexpected events. In this sense, we can compare the natural echo-system – made up of a complex network of interactions, integrations, retro-actions that make it mostly unforeseeable and unpredictable – with the electronic system. On a modern aircraft, the electronic echo-system is mostly not known by its user. Pilots only know which button they have to press, what the probable outcome is, and ignore what lies is in between. It is a new way of operating that has pros and cons. To assess the real impact on safety, we ought to analyze the relationship between man and technology, by shifting from HMI (Human Machine Interaction) to HCI (Human Computer Interaction). A further step will lead to Human Machine Engineering and/or Human Machine Design (Boy, 2011). The core of this approach is that the focus of the research should be on human-centred design; in other words, the final user – with his/her mental patterns activated in real scenarios – should be regarded as the core of the entire project for a new form of automation.

From an engineering perspective, it is very strange that concepts routinely used to describe the role of onboard automation miss a basic focal point: the final user. In fact, when we talk about technology, we refer to bolt-on versus built-in systems. These expressions indicate the different pattern of integration of onboard technology. Bolt-on indicates the introduction of a new technology on board an airplane conceived without automation. It is a reactive mode that strives to combine the old engineering philosophy with new devices. It is a kind of "patchwork" which requires several local adjustments to combine new requirements with old capabilities. The expression "built-in" indicates the development of a new technology incorporated in the original project. Every function is integrated with the airplane's systems, making every action consistent with the original philosophy of use. The paradox is that the final user is left out of the original project. Nowadays, after several avoidable accidents, pilots are involved in the early stage of design in order to produce a user-friendly airplane.

6. Why automation?

Two main reasons led to the decision to adopt onboard automation: the elimination of human error and economic aspects. The first element stems from the general view whereby human performance is regarded as a threat to safety. As such a topic would require a paper in itself, it is more appropriate to briefly mention some references for students eager to investigate the topic thoroughly. The second element is easier to tackle since we can even quantify the real savings related to, say, lower fuel consumption. According to IATA estimates, "a one percent reduction in fuel consumption translates into annual savings amounting to 100,000,000 dollars a year for IATA carriers of a particular State". (ICAO, 1998). Aside from fuel, the evolution of onboard technology over the years has led to a dramatic improvement in safety, operational costs, workload reduction, job satisfaction, and so forth. The introduction of the glass cockpit concept allows airlines to reduce maintenance and overhauling costs, improve operational capabilities and ensure higher flexibility in pilot training.

a.   Fuel consumption

A crucial item in an airline's balance sheet is the fuel cost. Saving on fuel is vital to remain competitive on the market. The introduction of the "fly-by-wire" concept helps to reduce fuel consumption in at least three areas: weight, balance and data predictions.

1.   The fly-by-wire concept has brought a tangible innovation. Inputs coming from the pilots' control stick are no longer conveyed via cables and rods directly to the aerodynamic surfaces. In fact, the side-stick (or other devices designed to meet pilots' demands regarding a conventional yoke) provides input to a computer which – via optic fibres – sends a message to another computer placed near the ailerons or stabilizer. This computer provides input to a servo-mechanism to move the surfaces. Therefore, there is no longer any need for steel cables running through the fuselage and other weighty devices such as rods, wheels, etc. This also significantly reduces the aircraft's weight and improves fuel consumption, since less power is required to generate the required lift.

2.   The second area that contributes to saving fuel is aircraft balance. The aircraft must be balanced to maintain longitudinal stability (pitch axis in equilibrium). This equilibrium may be stable, unstable or neutral. In a stable aircraft, the weight is concentrated in front of the mean aerodynamic chord. Basically, this means that the stabilizer (the tail) should "push-down" (or, technically, induce de-lifting) to compensate for the wing movements. In an unstable equilibrium, the balance of weight is shifted sensibly backwards compared to a stable aircraft. In other words, the stabilizer should generate lift to compensate for the wing movements. Where does the problem lie with unstable aircrafts? A stable aircraft tends to return to its original state of equilibrium after it deviates from the latter, but is less

manoeuvrable since the excursion of the stabilizer is narrower. On the contrary, unstable equilibrium causes an increasingly greater magnitude of oscillations as it deviates from the initial point. It makes the aircraft more manoeuvrable but unstable. In practical terms, in these kinds of airplanes pilots are required to make continuous corrections in order to keep the aircraft steady. This is why the computer was introduced to stabilize the airplane with continuous micro-corrections. This significantly reduces the pilots' workload for flying smoothly. Moreover, due to the distribution of weight concentrated on the mean aerodynamic chord, an unstable aircraft consumes less fuel.

3. The third factor helping pilots is a database capable of computing in real time any variation to the flight plan either on the horizontal path (alternative routes, short-cut, mileage calculations, etc.) or on the vertical profile (optimum altitude, top of descent to manage a low drag approach, best consumption speed, and so forth). This enhances the crew's decision-making task in choosing the best option in order to save fuel.

b. Maintenance costs

The glass cockpit concept enables airlines to reduce maintenance and overhauling costs. In conventional airplanes, every instrument had its box and spare part in the hangar. Whenever a malfunction was reported by the crew, maintenance personnel on the ground fixed it by replacing the apparatus or swapping the devices. All these actions required a new component for every instrument. If we consider that an airliner has roughly one million spare parts, we can easily understand the economic breakthrough offered by the glass cockpit concept.

In these airplanes, a single computer gives inputs to several displays or instruments. The maintenance approach is to change a single computer rather than every component or actuator. Based to this operating method, few spare parts are required in the hangar: no more altimeters, no more speed indicators, no more navigation displays (often supplied different manufacturers). Moreover, training of maintenance personnel is simplified as it focuses on a few items only which, in turn, allows for increased personnel specialisation.

c. Selection and Training costs

The fast growth pace of the airline industry over the last decades has generated concern about the replacement of older pilots, since training centres cannot provide the necessary output for airline requirements. Hiring pilots from a limited base of skilled workers creates a bottleneck in the industrial supply of such an essential organizational factor as are pilots for an airline. Automation has facilitated the hiring of new pilots, since the basic skills are no longer a crucial item to be verified in the initial phase of a pilot's career. If, barely thirty years ago, it would not have been sufficient "to have walked on the moon to be hired by a major airline", as an expert pilot ironically put it, nowadays the number of would-be pilots has increased exponentially. In the current industrial philosophy, almost everyone would be able to fly a large airliner safely with a short amount of training. This phenomenon gives rise to new and urgent problems, as we'll now see.

Besides the selection advantages, broadening the potential pilot base enables airlines to save money for the recurrent training of pilots or to reduce transition costs. Indeed, once the manufacturer sets up a standardized cockpit display, the latter is then applied to a series of aircraft. If we look at the Airbus series comprising A-319, A-320, A-321, A-330, A-340, etc., we realize how easy is to switch from one airplane to another. In this case, the transition course costs are significantly lower since pilots require fewer lessons; indeed, aside from certain specific details, the only difference involves the performance (take-off weight, cruise speed, landing distances, etc.).

Since pilot training costs make up a considerable portion of an airline's budget, it is financially convenient to purchase a uniform fleet made up of same "family" of airplanes. Often, the regulations enable airlines to use a pilot on more than one airplane belonging to a "family". Consequently, this leads to shorter transition courses, operational flexibility as pilots get to fly several aircrafts at a time, and in the long-run, better standardization among pilots.

Some authors have pointed out how automation has redefined the need for different training processes and crew interaction (Dekker, 2000).

d.   Operational flexibility

A pilot flying with no aids at all, be they mechanical, electrical or electronic, is limited in many ways.

He/she must fly at low altitude, because of his/her physiological limits (hypoxia), he/she cannot fly too fast since the effort on the yoke exceeds his/her physical power, he/she cannot even fly in bad weather (clouds or poor visibility) since he/she must maintain visual contact with the ground. Automation allows for overcoming such limitations. Higher flight levels also mean lower fuel consumption and the possibility of flying out of clouds. Faster speeds allow for reaching the destination earlier and completing multiple flights a day. Onboard instruments enable pilots to achieve better performance. Let's imagine an approach in low visibility conditions. In the beginnings, pilots would rely on a Non-Directional Beacon (NDB) as an aid to find the final track to land on the runway. Safety measures were implemented such as the operating minima. These implied that during a final landing approach, the pilot had to identify the runway before reaching a certain altitude (landing minima). Obviously, as instruments became more reliable, the landing minima were lowered. Subsequently, the introduction of the VOR (Very High Frequency Omnidirectional Range), which provided a more accurate signal, allowed for performing an approach closer to the runway and at a lower "decision altitude", as the safety margins were assured. When the ILS (Instrumental Landing System) was introduced on board, pilots could also rely on vertical profile indications. This implied higher safety margins that led the regulatory bodies to once again lower the landing minima. As the ILS became increasingly precise and accurate, the landing minima were lowered till they reached the value of zero. This means that an airplane can touch down with such low visibility that pilots must rely on autopilot to perform such an approach. This is due to human limitations such as visual and vestibular illusions (white-out, wall of fog, duck under, etc.) that may impair the pilot's performance. Let's imagine an approach during the '50s: with a visibility of 1,000 metres at the destination airport, the airplane should have diverted to another airport because the pilots would not have been able to attain the airport visual reference at the decision altitude (landing minima). Nowadays, the same airport could operate with 100 metres visibility due to improved transmitting apparatus on the ground and flight automation that enhances pilot performance.

A problem evidenced by several authors concerns the shift in responsibility of the people managing the automation system. With the performances brought about by automation that exceed human capabilities, a failure in the automatic system places pilots in a situation in which they have no resources available for coping with an unexpected event. In one case, where the pilots were performing a low visibility approach, the automatic system went out of control at very low altitude, causing the airplane to pitch down and hit hard on the runway. Experts acknowledge that the pilots' recovery in that case was beyond reasonable intervention, that is to say "impossible" (Dismukes, Berman, Loukopoulos, 2008). Indeed, the time available for detecting, understanding and intervening was so short that it was virtually impossible to solve the airplane's faulty behaviour. Who is responsible in such a case? Is it correct to refer to the pilots' mismanagement, poor skill, or untimely reaction when they are operating outside their safety boundaries? If the operations are conducted beyond human capabilities we should also review the concept of responsibility when coping with automation. As Boy and Grote put it: *"Human control over technical systems, including transparency, predictability, and sufficient means of influencing the systems, is considered to be the main prerequisite for responsibility and accountability"* (Boy, 2011).

Classic T-model

7. Different cockpit presentation

In recent years, cockpit design has undergone somewhat of a revolution. Old aircraft were designed to satisfy pilots' needs and relied upon slight modifications in the previous onboard scanning pattern. A common standardized cockpit display emerged in the early '50s: the T-model. It included the basic instruments such as artificial horizon (in the top-centre), anemometer (speed indicator) on the left, the altimeter on the right and compass (indicating the heading and track) in the bottom-centre position. Two further instruments, the side-slip indicator and vertical speed indicator were added later on. The adjacent picture shows the classic T-model.

Primary flight Display

With the introduction of the glass cockpit, the traditional flight instrument display was replaced by a different presentation encompassing more information, greater flexibility, colour coding and marking, but at the same time, it could lead to information overload, as several parameters are displayed in a compact area. In a single display, known as the PFD (Primary Flight Display), multiple information is included not only for basic parameters, but also for navigation functions, approach facilities, automatic flight feedback, flight mode awareness, and so forth.

Since the pilot needs to cross-check a great number of instruments at a glance, the design of traditional instruments was developed according to a pattern of attention, from the more important information to the more marginal information. Each instrument had its own case and its functional dynamics was perfectly known by the pilot; it was easy to detect, easy to use during normal operations and easy to handle in case of failure. The switch positions and shapes were positioned on board according to a pattern of use and were instantly recognisable at the touch. It was common for pilots to perform the checklist according to the "touch and feel" principle, by detecting whether a switch was in the required configuration by confirming its on/off position. Almost every switch had a peculiar shape: i.e., rough and big for operative ones, smooth and small for less important switches. A form of training implemented in most flying schools was the "blind panel" exercise, which consisted in covering the trainee's eye with a bandage and asking him/her to activate the switch prompted by the instructor. Being familiar with the physical ergonomics of the cockpit proved very helpful. The pilot knew how much strength was required to activate a command, how much the arm had to be stretched to reach a knob and so forth. This exercise enhanced the pilot's skill, or the skill level of operation (according to Rasmussen's Skill-

Rule-Knowledge paradigms). Regarding the input to the flight controls, it was assured via flight control wheels and sticks, cables and rods, in order to enable the pilot to intervene directly on the aerodynamic surfaces such as the ailerons, rudder and stabilizer.

With regard to displays, the traditional instrument configuration – made up of drum pointers, single instrument boxes and "touch and feel" overhead panel – was replaced by a new approach towards onboard information available to the pilot. The "touch and feel" philosophy was replaced by the "dark cockpit" approach. A dark, flat, overhead panel was adopted to show the pilot that everything was OK. Failures or anomalies were detected by an illuminated button, recalled by a master caution/warning light just in front of the pilots' eyes. In the unlikely (but not impossible) situation of smoke in the cockpit severely impairing the pilots' sight, (due to the concentration of a thick layer of dense, white fog), pilots of conventional airplanes could retrieve the intended system's configuration "by heart", by detecting the switch position with a fingertip.

Vice-versa, let's imagine a dark panel with smoke on board. It is truly challenging to spot where to place your hand and, above all, what the system's feedback is, since the system configuration is not given by the switch position but by an ON/OFF light, which is often undetectable.

8. Ergonomics

The problem with innovation, especially in a critical context for safety as is aviation, lies in the interaction between a community of practice and a new concept, conceived and implemented by engineers. This implies a relationship between automation and ergonomics. Ergonomics is a word deriving from the Greek words *ergon* (work) and *nomos* (law), and it is a field of study aimed at improving work conditions so as to guarantee optimal adaptation of the worker to his/her environment. We may identify three main types of ergonomics: physical, cognitive and social (or organizational).

Initially, with the onset of the first generation of automation (mechanical), ergonomists studied physical ergonomics, namely: how to reach a control, how much force is required to operate a lever, visibility of displayed information , seat position design, and so forth. For example, several accidents occurred due to misuse of the flap lever and landing gear lever. Indeed, these were positioned near one another and had similar shapes, leading the pilots to mistake them (Koonce, 2002). The ergonomists found a straightforward and brilliant solution: they attached a little round-shaped rubber wheel to the lever tip to indicate the landing gear, and a wing-shaped plastic cover indicating the flaps. Moreover, the two levers were separated in order to avoid any misuse.

In the next generation of automation, namely electrical automation, the ergonomists aimed to improve the cockpit design standardization. A pilot flying on an airplane had difficulty in forming a mental image of the various levers, knobs and displays, as every manufacturer arranged the cockpit according to different criteria. Dekker has clearly illustrated a case of airplane standardization: the position of the propeller, engine thrust and carburettor in a cockpit (Dekker, 2006). Pilots transitioning to other aircrafts struggled to remember the position of every lever since they were adjacent to each other, and the possibility of mistaking them was very high. Moreover, during the training process there was a risk of the so-called "negative transfer", that is, the incorrect application of a procedure no longer suited to the new context. For example, the switches on Boeing aircrafts have top-down activation, while Airbus has chosen the swept-on system (bottom-up). A pilot transitioning from a Boeing 737 to an Airbus A-320 may quickly learn how to switch on the systems, but in conditions of stress, fast-paced rhythm and high workload, he/she may return to old habits and implement them incorrectly. Social ergonomics tries to eliminate such difficulties.

Cognitive ergonomics studies the adaptability of the technology to the mental patterns of the operator. The third generation of automation introduced on board aircrafts gave rise to many concerns about whether the instrument logic was suitable for being correctly used by pilots. From a cognitive perspective, the new system should be designed according to the user's need, while bearing in mind that pilot-friendly is not equivalent to user-friendly (Chialastri, 2010). A generic user is not supposed to fly, as some engineers erroneously tend to believe. Pilots belong to a professional community that share a mental pattern when coping with critical operational situations. The *modus operandi* is the result of a life-long in-flight experience. A professional community such as the pilot community is target-oriented, and aims at the final goal rather than observing all the required procedural tasks step-by-step. This is why we talk about "shortcuts", "heuristics", and so forth.

For example purposes, let's discuss the introduction of a new airplane with variable wings: the F-111. It was conceived by the designers with a lever in the cockpit to modulate the wings from straight (useful at low speed) to swept (to fly at high speed). In the mind of the designer (who is not a pilot), it seemed perfectly sensible to associate the forward movement of the lever in the cockpit with the forward movement of the wings: lever forward – straight wings, lever back – swept wings. When the airplane was introduced in flight operations, something "strange" occurred. Pilots associated the forward movement with the concept of speed. So, when they wanted high speed, they pushed the thrust levers forward, the yoke down to dive and, consequently, the wing-lever forward (incorrectly so). For a thorough analysis of this case, see Dekker (2006).

All this occurred due to the confusion between generic users and specialized users, as are pilots. As Parasumaran puts it: "Automation does not simply supplant human activity but rather changes it, often in ways unintended and unanticipated by the designers of automation and, as a result, poses new coordination demands on the human operator" (Parasumaran, Sheridan, 2000).

9. Automation surprise

The interaction between pilots and technologies on board the aircraft raises some concerns regarding an acknowledged problem: the automation surprise. This occurs when pilots no longer know what the system is doing, why it is doing what it does and what it will do next. It is an awful sensation for a pilot to feel to be lagging behind the airplane, as – ever since the early stages of flight school – every pilot is taught that he/she should be "five minutes ahead of the airplane" in order to manage the rapid changes occurring during the final part of the flight. When approaching the runway, the airplane's configuration demands a higher workload, the communication flow with air traffic controllers increases and the proximity to the ground absorbs much of the pilot's attention.

Once all the fast-pace activities are handled with the aid of automation, the pilot may shed some of the workload. When automation fails or behaves in a "strange" manner, the workload increases exponentially. This is only a matter of the effort required to cope with automation, and could be solved with additional training. A more complicated issue – arising from investigations on some accidents – is the difference between two concepts, namely: "situational surprise" and "fundamental surprise" (Wears, 2011). To summarise our explanation, we may refer to the example mentioned by Wears regarding an everyday situation: a wife returns home unexpectedly and finds her husband in bed with the waitress. She is shocked and the only thing she can say is: "Darling, I am surprised". Her husband calmly replies, "no, darling, I am surprised; you are astonished". In this example, the husband knows he is doing something regrettable, but accepts the risk even if he doesn't expect his wife to return home. He is surprised, but has calculated even the worst case (wife back home sooner than expected). The wife comes home and finds a situation she is not prepared for. She is surprised and doesn't know what to do at first, because she lacks a plan. No doubt she will find one...

Therefore, situation surprise arises when something occurs in a pre-defined context, in which the pilots know how to manage the situation. Such a situation is unexpected, being an abnormal condition, but nonetheless remains within the realm of the pilot's knowledge. Moreover, it returns to being a manageable condition once the pilot acknowledges the difference between a normal and an abnormal condition, by applying the relevant abnormal procedure. After the initial surprise due to the unexpected event, the pilot returns to a familiar pattern of operation acquired with training.

A so-called fundamental surprise is a very different circumstance: it implies a thorough re-evaluation of the situation starting from the basic assumptions. It is an entirely new situation for which there is no quick-fix "recipe" and is difficult to assess, since it is unexampled. The pilots literally do not know what to do because they lack any cognitive map of the given situation. This concept can be well summarised by the findings following an accident involving a new generation airplane. When analysing the black box after an accident, it is not rare to hear comments such as: "I don't know what's happening" – this is a something of a pilot's nightmare.

It should be noted that some failures occurring on board highly automated aircrafts are transient and are very difficult to reproduce or even diagnose on the ground. Even the manufacturer's statements regarding electrical failures are self-explaining, as a failure may produce different effects in each case or appear in different configurations depending on the flight phase, airplane configuration, speed and so forth. Sometimes, an electrical transient (also influenced by Portable Electrical Devices activated by unaware passengers) could trigger certain airplane reactions that leave no memory of the root cause.

10. Automation issues

Although automation has improved the overall safety level in aviation, some problems tend to emerge from a new way of operating. A special commission was set up in 1985 by the Society of Automotive Engineers to determine the pros and cons arising from flight deck automation. Nine categories were identified: situation awareness, automation complacency, automation intimidation, captain's command authority, crew interface design, pilot selection, training and procedures, the role of pilots in automated aircrafts (ICAO, 1998). Sometimes, pilots loose situation awareness because they lag behind the automation logic. A pilot should always know what the automation system is going to do in the next five minutes, to either detect any anomalies or take control of the airplane. In fast-paced situations – such as crowded skies, proximity with the terrain and instrumental procedures carried out in poor visibility – a high number of parameters must be monitored. As the workload increases, the pilot tends to delegate to the automation system a series of functions in order to minimise the workload. It is important to specify that the physical workload (the number of actions performed within a given time frame) differs from the cognitive workload in that the latter implies a thorough monitoring, understanding and evaluation of the data coming from the automation system. The paradox involving airplane automation is that it works as an amplifier: with low workloads, it could lead to complacency ("let the automation system do it") that reduces alertness and awareness, while the latter increase with high workloads, due to the high number of interactions and data involved in fast-paced situations. Plainly said: "When good, better; when bad, worse".

Poor user interface design is another issue. Norman, Billings et al. have studied the way humans interact with automation and have developed cognitive ergonomics. There are several studies concerning the optimal presentation (displays, alarms, indicators, etc.) for pilot use. For example, old displays were designed in such a way that each single instrument provided data for a specific domain only: the anemometer for speed, the altimeter for altitude and flight levels, and so forth. With the integrated instruments, all the required data is available at a glance in a single instrument, which not only provides colour-coded information but also on-demand information by means of pop-up features, so that it can be concealed during normal operations and recalled when needed.

Several issues are related to displays, but only two are addressed here: cognitive mapping of instruments and human limitations in applying the colour-code to perception.

We have to consider that the replacement of round-dial black and white instruments, such as the old anemometer, with coloured multi-function CRT (Cathodic Ray Tube) instruments, has determined a new approach by pilots in mentally analysing speed data. Indeed, with round-dial indicators, pilots knew at a glance in which speed area they were flying: danger area, safe operations or limit speed. Although it is not important to know the exact speed, according to a Gestalt theory concept we need to see the forest rather than every single tree.

Furthermore, from a cognitive perspective, we tend to perceive the configuration as a multiple instrument: for example, during the approach phase, the anemometer needle will point to 6 o'clock, the vertical speed indicator to 8 o'clock and so forth.

In the new speed indicator, the values appear on a vertical speed tape rolling up and down, with the actual speed magnified for greater clarity. Moreover, additional indicators are displayed on the speed tape, such as minimum speed, flap retraction speed and maximum speed for every configuration. In many ways, this system facilitates the pilot's task. Problems arise whenever failures occur (electrical failure, unreliable indication etc.), as the markings disappear.

The situation worsens compared to the old indicators, as pilots needs to create a mental image of the speed field in which they are flying. For example, it is unimportant for us to know whether we are flying at 245 or 247 knots, but it becomes important to know that we are flying at 145 knots rather than 245 knots. This means that the mental mapping of the speed indication involves a higher workload, is time-consuming and also energy-consuming. Let's take a look at these two different speed indicators to grasp the concept.

The second issue is related to human performance and limitations. The human eye may detect colours, shape, light and movement with foveal vision. There are two kinds of receptors in the fovea: the cones and the rods.

The cones are responsible for detecting shapes and colours, in a roughly 1 cent large area, and account for focused vision. Instead, rods are responsible for peripheral vision and are able to detect movement and light. This means that colour-coded information should appear in an area just in front of the pilot's head, as only the cones may such information. In fact, since an input appearing to the side can be detected by the rods, which are insensitive to colour and shape, the pilot will not notice it unless he stares at it directly.

Other relevant indicators for pilots are symmetry and context. The needle in a round-dial instrument satisfies both these requirements, since we may easily detect which field of operation we are currently in (near the upper limit, in the middle, in the lower margin), in addition to the trend (fast-moving, erratic, gradual) and symmetry with nearby parameters. A digital indicator shows this information in an alternative way, by arranging numbers on a digital display, as can be clearly seen in the above picture. The configuration on the left can be grasped at first glance, while the second set of information requires a considerable effort to detect differences.

Disrupted symmetry   **Digital information**

Another area worthy of attention is the greater magnitude of the errors made by pilots when flying highly automated airplanes. Pilots flying older aircrafts normally relied on their ability to obtain the required data using a heuristic and rule-of-thumb approach. This method was not extremely accurate but roughly precise. Nowadays, with the all-round presence of computers on board, flight data can be processed very precisely but with the risk of gross errors due to inaccurate entries made by pilots. An example taken from everyday life may help to explain the situation: when setting older-version alarm clocks, the user's error might have been limited to within the range of a few minutes, whereas the new digital alarm clocks are very accurate but subject to gross errors – for example 8 PM could be mistaken for 8AM.

Lower situation awareness means that pilots no longer have the big picture of the available data. The aviation truism, "trash in-trash out", is applicable to this new technology as well. Having lost the habit of looking for the "frame", namely the context in which the automation system's computed data should appear, pilots risk losing the big picture and, eventually, situation awareness. Unsurprisingly, there have been cases of accidents caused by inconsistent data provided by the computer and uncritically accepted by pilots. This is neither a strange nor uncommon occurrence. Fatigue, distractions, and heavy workload may cause pilots to lower their attention threshold. In one case, a pilot on a long haul flight entered the available parameters into the computer to obtain the take-off performance (maximum weight allowed by the runway tables, speed and flap setting). He mistook the take-off weight with other data and entered the parameters into the computer. The end result was a wrong take-off weight, wrong speed and wrong flap setting. The aircraft eventually overrun the runway and crashed a few hundred meters beyond the airport fences.

Another issue linked the introduction of onboard automation is a weakening of the hierarchy implied by the different way of using the automation system. The captain is indeed the person with the greatest responsibility for the flight and this implies a hierarchical order to establish who has the final word on board. This hierarchical order is also accompanied by functional task sharing, whereby on every flight there is a pilot flying and a pilot not flying (or monitoring pilot). Airplanes with a high level of automation require task sharing, in which the pilot flying has a great degree of autonomy in programming the Flight Management System, in deciding the intended flight path and type of approach. Above all, the pilot flying also determines the timing of the collaboration offered by the pilot not flying, even in an emergency situation. In fact, since the crew must act in a procedural way in fulfilling the demands of the automation system, both pilots must cooperate in a more horizontal way compared to the past. The hierarchical relationship between captain and co-pilot is known as the "trans-authority gradient" (Hawkins, 1987) and, from a human factor perspective, should not be too flat nor too steep.

11. Case studies

Currently, several case studies could be cited to illustrate the relationship between pilots and automation. In the recent past, accidents have occurred due to lack of mode awareness (A-320 on Mount St. Odile), automation misuse (Delhi, 1999), loss of braking leading to loss of control (S. Paulo Garulhos, 2006), loss of control during approach (B-737 in Amsterdam, 2008), and several other cases.

From these accidents, we chose a couple of peculiar cases. The first case occurred in 1991 and involved an Airbus flying at cruise altitude. While the captain was in the passenger's cabin, the co-pilot tried to "play" with the FMS in order to learn hands-on. He deselected some radio-aids (VOR: Very High Frequency Omni-directional Range) from the planned route. After the tenth VOR was deselected, the airplane de-pressurized, forcing the pilot to perform an emergency descent. From an engineering perspective there was a bug in the system but, moreover, it was inconceivable for the pilot to link the VOR deselection (a navigation function) to the pressurization system.

The second case is related to a B-727 performing a low-visibility approach to Denver. The flight was uneventful till the final phase. The sky was clear but the city was covered by a layer of fog that reduced the horizontal visibility to 350 ft. and the vertical visibility to 500 ft. The captain was the pilot flying, in accordance with the company regulations, and used the autopilot as specified in the operating manual. The chosen type of approach was an ILS Cat II. The ILS is a ground based navigation aid that emits a horizontal signal to guide the airplane to the centre of the runway together with a vertical signal to guide the aircraft along a certain slope (usually 3°) in order to cross the runway threshold at a height of 50 ft. The regulations state that once the aircraft is at 100 ft., the crew must acquire the visual references to positively identify the runway lights (or markings). If not, a go-around is mandatory. Once the captain has identified the runway lights he must proceed to land manually, by switching off the autopilot at very low altitude. There are certain risks associated with this manoeuvre, since – with impaired visibility – the pilot loses the horizon, depth perception and the vertical speed sensation, and could be subject to spatial disorientation due to the consequences of the so-called "white-out" effect. In this case, however, nothing similar occurred. The autopilot duly followed the ILS signals, but due to a random signal emitted in the last 200 ft., the aircraft pitched down abruptly. The captain tried to take over the controls, though unsuccessfully. He later reported – during the investigation – that the window was suddenly "full of lights", meaning that the aircraft assumed a very nose-down pitch attitude. The high rate of descent, coupled with the surprise factor and low visibility, did not allow the captain to recover such a degraded situation: the airplane touched down so hard that it veered off the runway, irreversibly damaging the fuselage.

This case emphasizes the relationship between pilots and liability. Operating beyond human capabilities, the pilot finds himself/herself in a no-man's land where he/she is held responsible even when he/she cannot sensibly regain control of the airplane. As stated by Dismukes and Berman, it is hardly possible to recover from such a situation.

12. Conclusion

This brief paper highlights certain aspects concerning the introduction of automation on board aircrafts. Automation has undeniably led to an improvement in flight safety. Nevertheless, to enhance its ability to assure due and consistent help to pilots, automation itself should be investigated more thoroughly to determine whether it is suitable in terms of human capabilities and limitations, ergonomics, cognitive suitability and instrument standardization, in order to gradually improve performance.

13. References

Air Pilot Manual, (2011), *Human Factors and pilot performance*, Pooleys, England

Alderson D.L., Doyle J.C., "Contrasting view of Complexity and Their Implications For Network-Centric Infrastructures", in *IEEE Transaction on Systems, Man and Cybernetics* – part A: Systems and Humans, vol 40 No. 4, July 2010

Amalberti R., "The paradox of the ultra-safe systems", in Flight Safety Australia, September-October 2000

Bagnara S., Pozzi S., (2008). Fondamenti, Storia e Tendenze dell'HCI. In A. Soro (Ed.), *Human Computer Interaction. Fondamenti e Prospettive*. Monza, Italy: Polimetrica International Scientific Publisher.

Lisanne Bainbridge, (1987), "The Ironies of Automation", in "New technology and human error", J. Rasmussen, K. Duncan and J. Leplat, Eds. London, UK: Wiley,.

Barnes C., Elliott L.R., Coovert M.D., Harville D., (2004), "Effects of Fatigue on Simulation-based Team Decision Making Performance", Ergometrika volume 4, Brooks City-Base, San Antonio TX

Boy G., a cura di, (2011), *The Handbook of human Machine Interface – A Human-centered design approach*, Ashgate, Surrey, England

Chialastri Antonio (2011), "Human-centred design in aviation", in *Proceedings of the Fourth Workshop on Human Centered Processes*, Genova, February 10-11

Chialastri Antonio (2011), "Resilience and Ergonomics in Aviation", in *Proceedings of the fourth Resilience Engineering Symposium* June 8-10, 2011, Mines ParisTech, Paris

Chialastri Antonio (2010), "Virtual Room: a case study in the training of pilots", HCI aero-conference, Cape Canaveral

Cooper, G.E., White, M.D., & Lauber, J.K. (Eds.) (1980) "Resource management on the flightdeck," Proceedings of a NASA/Industry Workshop (NASA CP-2120)

Dekker S., "Sharing the Burden of Flight Deck Automation Training", in The International Journal of Aviation Psychology, *10*(4), 317–326 Copyright © 2000, Lawrence Erlbaum Associates, Inc.

Dekker S. (2003), "Human Factor in aviation: a natural history", Technical Report 02 - Lund University School of Aviation

Dekker S., Why We need new accident models, Lund University School of Aviation,Technical Report 2005-02

Dismukes, Berman, Loukopoulos (2008), *The limits of expertise*, Ashgate, Aldershot, Hampshire

Ferlazzo F. (2005), *Metodi in ergonomia cognitiva*, Carocci, Roma

Flight Safety Foundation (2003), "The Human Factors Implications for Flight Safety of Recent Development In the Airline Industry", in *Flight Safety Digest*, March-April

Hawkins Frank (1987), *Human factor in Flight*, Ashgate, Aldershot Hampshire

Hollnagel E., Woods D., Leveson N. (2006) (a cura di), *Resilience Engineering – Concepts and Precepts*, Ashgate, Aldershot Hampshire

Hollnagel Erik, (2008) "Critical Information Infrastructures : should models represent structures or functions ?", in *Computer Safety, Reliability and Security*, Springer, Heidelberg

Hollnagel Erik, (2009), *The ETTO Principle – Efficiency-Thoroughness Trade-Off*, Ashgate, Surrey, England

Hutchins Edwin (1995), "How a cockpit remembers its speeds", Cognitive Science, n. 19, pp. 265-288.

IATA (1994), *Aircraft Automation Report*, Safety Advisory Sub-Committee and Maintenance Advisory Sub-committee.

ICAO - Human Factors Digest No. 5, Operational Implications of Automation in Advanced Technology Flight Decks (Circular 234)

ICAO (1998) – Doc. 9683-AN/950, Montreal, Canada

Köhler Wolfgang (1967), "Gestalt psychology", Psychological Research, Vol. 31, n. 1, pp. 18-30

Koonce J.M., (2002), *Human Factors in the Training of Pilots*, Taylor & Francis, London

Maurino D., Salas E. (2010), *Human Factor in aviation*, Academic Press, Elsevier, MA, USA

Morin E. (2008), *Il Metodo – Le idee: habitat, vita organizzazione, usi e costumi*, Raffaello Cortina editore, Milano

Morin E. (2004), *Il Metodo – La vita della vita*, Raffaello Cortina editore, Milano

Morin E. (2001), *Il Metodo – La natura della natura*, Raffaello Cortina editore, Milano

Morin E. (1989), *Il Metodo – La conoscenza della conoscenza*, Feltrinelli editore, Milano

David Navon (1977), "Forest before trees: the precedence of global features in visual perception", Cognitive Psychology, n. 9, pp. 353-383

Norman D. (1988), *The psychology of everyday things*, New York, NY: Basic Books, 1988.

Parasumaran, Sheridan , "A Model for Types and Levels of Human Interaction with Automation" IEEE Transactions on Systems, Man, and Cybernetics — part A: Systems and Humans, Vol. 30, No. 3, MAY 2000

Parasumaran R., Wickens C., "Humans: Still Vital After All These Years of Automation", in *Human Factors*, Vol. 50, No. 3, June 2008, pp. 511–520.

Ralli M. (1993), *Fattore umano ed operazioni di volo*, Libreria dell'orologio, Roma

Rasmussen J., *Skills, Rules, Knowledge: Signals, Sign and Symbol and Other Distinctions in Human Performance Models*, In "IEEE Transactions Systems, Man & Cybernetics", SMC-13, 1983

Reason James, (1990) *Human error*, Cambridge University Press, Cambridge

Reason James (2008), *The human contribution*, Ashgate, Farnham, England

Salas E., Maurino D., (2010), *Human factors in aviation*, Elsevier, Burlington, MA, USA

Wears Robert, Kendall Webb L. (2011), "Fundamental on Situational Surprise: a case study with Implications for Resilience" in *Proceedings of the fourth Symposium on Resilience Engineering*, Mines-Tech, Paris.

Woods D., Dekker S., Cook R., Johannesen L., Sarter N., (2010), *Behind human error*, Ashgate publishing, Aldershot, England

Woods D. D., *Modeling and predicting human error*, in J.Elkind S. Card J. Hochberg and B. Huey (Eds.), *Human performance models for computer aided engineering* (248-274), Academic Press 1990

Wright Peter, Pocock Steven and Fields Bob, (2002) "The Prescription and Practice of Work on the Flight Deck"*Department of Computer ScienceUniversity of YorkYork YO10 5DD*

6

Virtual Commissioning of Automated Systems

Zheng Liu[1], Nico Suchold[2] and Christian Diedrich[2]
[1]Otto-von-Guericke University, Magdeburg,
[2]Institut für Automation und Kommunikation (ifak), Magdeburg,
Germany

1. Introduction

Concepts for the digitalization of products and all the production related tasks in the manufacturing and process industry have been developed for several decades. The development began with the introduction of computer-based 2D design (Figure 1).

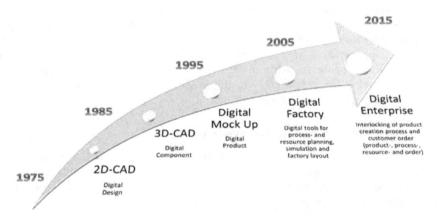

Fig. 1. Development stages for the Digital Factory (Bär, 2004)

Because of the technological developments in the information technologies field, more complex tasks in the product development process can be processed digitally (Brac, 2002). This belongs to the scope of the digital factory which includes the digitalization of models of the products and their integration into the manufacturing process chain. In (VDI 4499, 2008) the digital factory was defined as a "generic term for a comprehensive network of digital models and methods, including simulation and 3D visualization. Its purpose is the integrated planning, implementation, control and on-going improvement of all important factory processes and resources related to the product."

One of the topics in the digital factory is the Virtual Commissioning. As we know, the commissioning of the automated system is an important phase in the engineering, which makes visible whether the systems and components are planned, designed, produced and installed correctly according to the user requirements. However, this phase has been known as time-consuming and cost-consuming in practice (section 2.1) and can be improved by Virtual Commissioning.

State of the art is that the simulation of the system components is frequently used and the simulation is used without integration of automation devices and components, for example without PLCs (Programmable Logic Controller). HiL (Hardware in the loop) technologies are used only for individual components. The concept of digital factory combines both technologies.

During Virtual Commissioning the real plant is replaced by a virtual model according to the concept of digital factory. But the questions are: what digital data is necessary for the virtual plant model, how is the virtual plant implemented and configured, and how can these processes be integrated into the engineering lifecycle of manufacturing systems? This chapter gives comprehensive answers to these questions.

2. The Virtual Commissioning

The basic idea of Virtual Commissioning is to connect a digital plant model with a real plant controller (e.g. PLC- or HMI-Human Machine Interface) so that engineers from different fields such as design process and control have a common model to work with together. Thus, for instance, the PLC program can be tested virtually before the physical implementation is finished. Furthermore, the general functionalities of the plant can be validated and finally tested in an earlier phase. In this section, the general workflow of a traditional engineering process is introduced and its disadvantage discussed. Thereafter the Virtual Commissioning, as a possibility to overcome these problems, is introduced and described in detail in the subsequent subsections.

2.1 General workflow in the engineering process

The engineering of manufacturing plants begins with the construction and the rough mechanic planning. In this phase the general structure (layout) of the manufacturing system as well as the assignment of the machines in different production phases are determined, which is based on the result of product development.

The detailed planning is done in the next step. The kinematic simulation for the offline programming (OLP) of the robots is carried out, therefore the robot programs can already be created without the real robots, e.g. for the generation of a trajectory free of collision with other devices. In the phase of mechanic detail planning, a detailed CAD-based 3D cell layout (M-CAD) is created. The production workflow of a single machine is to be defined and every mechanical component in the cell can also be selected and documented in the list of materials. The results of the mechanical detail planning are used as input for the subsequent electrical detail planning. At this step the electrical energy supply as well as the control signals for the machines and their interconnection with each other is defined. This can be called E-CAD and is documented in the circuit diagram. The last phase of the detail planning is the PLC programming.

The last step in the classic engineering process is the commissioning, which can be understood as a phase of preproduction. At this phase the functionality of every single component in a function group is examined. After that it follows the interconnection check in turn from the level of function groups to the level of stations until the whole plant is complete - the bottom-up principle. The result of the commissioning is a functional plant which is ready for the production.

2.2 Today's problem in the commissioning

The above mentioned approach of the engineering process can be seen as the water fall model. A single step in this model can only be executed one after the other. There is no iteration step back to the previous steps. As a result, the debugging takes place only at the phase of commissioning. The following two main problems are identified here.

Bugs in the software

(Zäh & Wünsch, 2005) describes the time consuming for commissioning and PLC-programming in a whole project. It can be seen that the commissioning takes up 15-20% of the overall project duration, of which up to 90% is for commissioning the industrial electrics and control systems (e.g. PLC). Within the commissioning, up to 70% of the time is incurred by software errors (Figure 2).

The debugging of the PLC-Software is usually time consuming and may cause hardware damage. Since the commissioning is usually carried out under extreme time pressure, then a delay of the project must be calculated with financial penalties. Therefore it is desirable that the software debugging be carried out in an earlier phase.

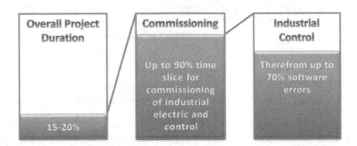

Fig. 2. Contribution of control software to project delay (Wünsch, 2008)

The error is detected too late

Many errors that are detected during the commissioning arise in the earlier phases. This is a big problem since the late detection of errors and the costs of their elimination can be very high (Figure 3). When an error occurs in the early phase, an overwork of the rest phases is not avoidable. For example, if a robot is wrongly placed in the 3D cell layout and this error is not detected until the commissioning phase, then it is a great effort with extra costs to eliminate this error because all the planning data related to this robot has already been modified.

Fig. 3. Costs incurred by the correction of errors, depending on the time of error detection (Kiefer, 2007)

2.3 State of the technology

One approach to overcome the disadvantages mentioned in subsection 2.2 is the full digital planning and simulation of the production line. Concepts, concrete process sequences and the necessary resources can be defined and approved already in the early planning phase. This activity belongs to the concept of digital factory.

Fig. 4. The engineering chain in the production with Virtual Commissioning

The focus of digital factory is the digital processing of product development and production planning based on the existing CAx data with a seamless workflow in PLM (product lifecycle management). Digital simulation methods and technologies are used to secure the planning results, to optimize the process and to respond more quickly to the changes than before. In this way the product qualities are improved. Another benefit is the reduction of the planning time as well as the overall project duration. The "Time-to-Market" is thus shortened which means another advantage with respect to economy.

To test the PLC-program before the real commissioning, the Virtual Commissioning can be applied (Figure 4). It serves as a smooth transition between digital and real factory. In the case of Virtual Commissioning, the physical plant, which consists of mechatronic components, is simulated with the virtual model. This simulated system is connected to the real controller (PLC) via simple connection or real industrial communication systems. The goal is to approximate the behavior of the simulated system to that of the real physical plant by connecting the commissioned PLC to the real plant without changes (Figure 5). Therefore the development and test of automation systems can be done parallel to the electric and mechatronic development. In the case of the real commissioning, the connection can be switched to the real system again.

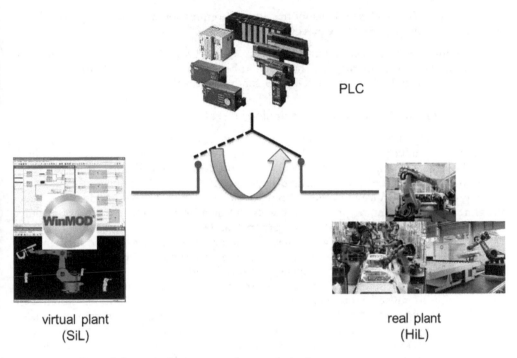

PLC

virtual plant
(SiL)

real plant
(HiL)

Fig. 5. From Virtual Commissioning to real commissioning

2.4 Component of the Virtual Commissioning

The virtualized plant, a mechatronic plant model, simulates the behavior of the real physical plant. It should respond correctly to the PLC control signals just as a real system. Besides, the whole process should be visualized for a better human observation. To fulfill these requests, a mechatronic model (Mühlhause et al., 2010) should be divided into a control-oriented behavior model and a kinematic 3D-model.

The behavior model simulates the uncontrolled behavior of the system. The behavioral states of the production resource are modeled and calculated by means of the logic and temporal components, which operate based on the control signals. In one PLC scan period, the behavior model should react to the control signals (output signals of PLC) according to the physical feature of the plant and give the feedback signal (input signals of PLC) back to the PLC – just as the behavior of a real system.

The kinematic 3D-model can be understood as a geometric model, which is based on the 3D-CAD model and enriched with additional information e.g. the grouping of the components which should be moved together and the definition of the DOF (degree of freedom) for every moving part. In contrast to today's mechanical CAD models, the mechatronic components do not just solely consist of the pure 3D CAD model, but also kinematic (including end positions) and electrical information like the electrical name of the respective device. For the purpose of synchronization, a signal coupling between the behavior model and kinematic 3D-model is needed.

The kinematic 3D-model is optional for the Virtual Commissioning, which helps for a better understanding by visualizing the movement behavior in 3D.

The communication between the mechatronic cell model and the real PLC can be realized via TCP/IP, field bus, Ethernet etc. There should be an interface to emulate the behavior of the communication. In the case of PROFIBUS, for example, a SIMBA-Card is needed.

2.5 Configuration of a Virtual Commissioning workstation

For the planning of a new production system, a layout plan for the necessary components (roles for the production resource to be used) and a rough workflow of the systems will be worked out based on a pattern (Figure 6). Based on this, a simulation model of the plant, including the geometric and kinematic design, will be built. This simulated layout is the starting point for the Virtual Commissioning, which is to be extended by the offline robot programming and planning process. At the step of detailed planning, the roles of resource are replaced by concrete units (function groups) under some fixed rules. For example, a certain combination of frequency inverter, the motor for a required electric moment of a torque, the type of clamp and valve cluster with the proper nominal pressure, or the robot and the weight of products to be operated, must be adhered to. These rules are to be observed by planners in the detailed planning. In a robot cell, for example, about 100 of such function groups must be matched and built. This function group can be stored in a kinematic library for the reusability. To build a simulation model, these previously created library elements can be added and connected with each other to describe the behavior of the corresponding component. Such a behavior simulation model has in average 30 inputs and outputs, which must be connected to the kinematics or to the PLC.

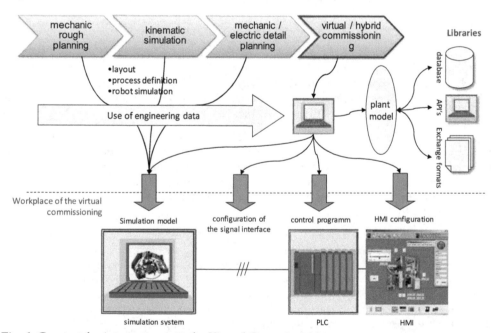

Fig. 6. Concept for integration into the Virtual Commissioning

After or in parallel to the layout planning and kinematic simulation, a suitable control program and operator visualization will also be developed, which run on a PLC or a PCS (Process Control System) system. The PLC program is developed in the following way: for each resource a corresponding control function block (Proxy FBs) is used, which is included in the control library. These Proxy FBs will be invoked in one or more SCF (Sequential Function Chart language IEC 61131-3). This control program will then be supplemented by plant-specific locks, a switch-off and parameterizations.

Afterwards, the PLC and the simulation systems are interconnected via the communication system or direct wiring. The I/O lists of the simulation system and the I/O lists of the control program must be properly connected with each other. Furthermore, each I/O has a number of features such as data type, range and unit, which need to be customized. The connection between the I/O of the Proxy FBs (or the communication configuration of the PLC) and the I/O lists of the simulation model must therefore be checked accordingly to the signal features.

A workstation for the Virtual Commissioning (Figure 6) consists of a simulation system on which the simulation model runs, a PLC which is usually connected with the simulation model via a bus system (e.g. PROFINET) and optionally a PCS.

In the phase of Virtual Commissioning, the plant which is to be tested is completely simulated in a digital model instead of a real one. This includes the automation equipment such as the clamps, robots and electric drives. When the PLC is connected to the real and digital systems at the same time, then it can be called the hybrid commissioning. For example, a robot with its controller can be integrated into a single system with other digital components, and the system should not detect any differences between the real and the digital components.

The current technology state assumes that the geometry and kinematics of the resources are provided by the manufacturer and as available libraries. In addition, the features of reused equipment may be stored in proprietary formats. These features and libraries are available in different storage systems (e.g. databases, repositories of tools and file systems).

Typical tools for Virtual Commissioning are Process Simulate/Process Simulate Commissioning and Delmia Robotics/Delmia Automation. In addition, the authors use the software WinMOD of the German company Mewes & Partner GmbH, which includes the component for 3D visualization. A circuit plan for electric planning can be edited for example with COMOS. There are also templates such as in Process Simulate, WinMOD, COMOS and Simatic S7, in which the corresponding features already exist in the library elements for converters, motors, robots etc. Today, the behavior models are not supplied by the resource producers. Instead, they are created by the users themselves for specific applications.

Besides the integration of a mechatronic plant model in the line-balancing scenario, a concept for the integration into the Virtual Commissioning was developed. The process of creating the workstation for the Virtual Commissioning has been explained above. The use of a mechatronic plant model should contribute to the construction of the workstation for Virtual Commissioning. The information which is extracted from existing data sources by using the mechatronic plant is able to represent the structure of the workstation (kinematic and behavioral simulation, PLC program, link-up between PLC and simulation model).

3. The mechatronic plant model with semantic techniques

The domain specific models in the individual life cycle phases usually have their own concepts depending on their physical phenomenon and tasks. This is essential because the models reflect only the necessary characteristics of the domains. A universal valid model is not thinkable. However, for seamless information flow it is necessary for the information consumers to have knowledge about the context of the transmitted information so that they can act in a proper way. (Rauprich et al., 2002) and (Wollschlaeger et al., 2009) point out that the historical separation of engineering disciplines, their workflows and thus the information integration is still an unsolved problem. One of the most common model structures in the manufacturing domain is the Product, Process Resource approach. It is the basis of digital planning of major tools like the Delmia tool set of Dassault Systemes (Kiefer, 2007).

Therefore, each piece of information has to be unambiguously identified and validated before it can be used in one of the engineering tasks. Additionally, its semantic including, its context information has to be specified and must be available. Mechatronic components, i.e. plant or machine parts consisting of mechanical, electrical and logical aspects, build up the modular blocks of the model, which are used in the engineering process. Thus these mechatronic components have to be represented by its semantic enriched information. Consequently, the mechatronic model must be based on semantic methods.

3.1 Mechatronic plant model

Today the information management in companies is characterized by a variety of available information, which is mostly stored distributed in a monolithic form and in different data sources. However ' the size of each data source increases continuously over time, so that a connection between the information can only be realized through significant human work. As a result, much of the available information is very difficult to find by the user. Hereby a situation relevant lack of information arises within the information stream in the company. The time required for the complex information search and retrieval rises exponentially with the increasing amount of information ' which results in significant costs. The mechatronic plant model is a logical layer above the design data and can be divided into the basic aspects of product, process and resource (Figure 7). Planning and other tasks in the plant life cycle can be executed by tool applications. These tools add new data to the existing data pool. These applications are modeled as so called facets (roles for the user data) in the mechatronic plant model. The resources could be further refined (see Figure 7). The basic idea of the model is to disperse all aspects (products, process and resource) in smaller features. For example, a construction group has the feature of the diameter of the pressure pipe connection or movement duration of a valve. The typical process features are process step number and number of the predecessor or successor step. If the mechatronic model is used for the acquisition of the data storage, the addressing data (e.g. data source type, URI of the data source) is also to be modeled. The mechatronic model is thus a hierarchical structure of sub-elements based on the three aspects product, process and resource. These three aspects are then to be further dissected into even smaller features, until no more division can be undertaken. In addition, these features are categorized so that semantic information about the individual elements of an aspect and its relationship with other elements of other aspects in the model can be enriched. (Figure 8)

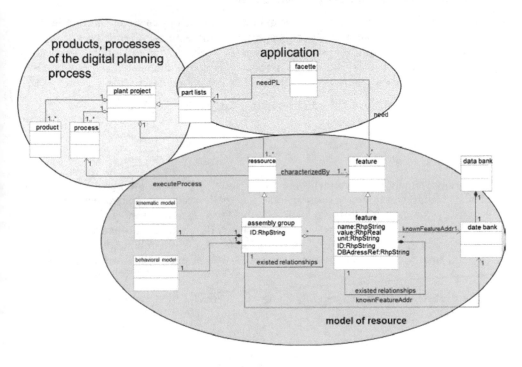

Fig. 7. Basic structure of the mechatronic plant model

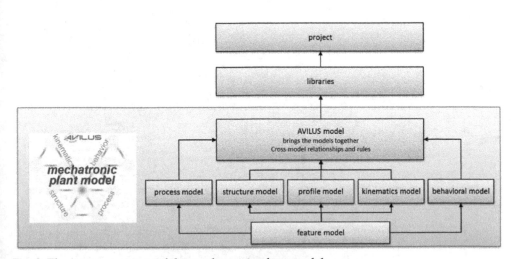

Fig. 8. The intern structure of the mechatronic plant model

3.2 Potential of the feature-based modeling approach

The basic idea of focusing on the features of engineering is to make engineering decisions and the underlying rules simpler to relationships between features. A unified set of features, for example coming from the standardization of technical product data, enables an automated execution of rules. This makes it possible to resolve the relationship with defined rules between feature carriers (e.g. valve or clamp or behavioral description and proxy FB).

This means the following assistance functions are available for the planner:

- Consistency check
- (semi-) automatic assignment
- Restrict the solution

An important element of the mechatronic plant model is the coupling between kinematics and behavior of the mechatronic components (aggregation relationship between the module as well as the kinematics and behavior model in Figure 7) and the integration of the PPR model (product, process and resource). Figure 7 shows the connection between resource and process (executeProcess). For the purpose of semantically unambiguous modeling, the description language OWL was used, which can clearly formulize the relations between the classes.

The extensive relationship between kinematics and behavior model are described in Figure 9.

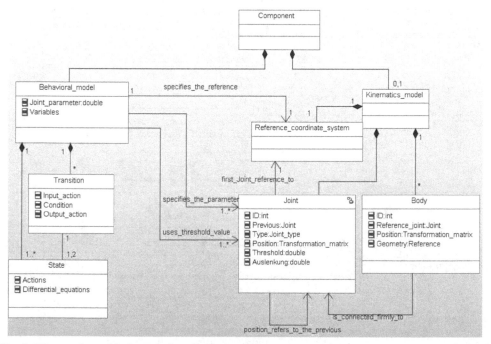

Fig. 9. Kinematic and behavioral model

In summary, the mechatronic plant model has the following essential basic elements:

- Basic structure is the grouping of product, process and resource
- Relationships are defined between the basic elements
- The PPR elements are then further specialized
- The properties of the PPR elements are described with features
- The features can be structured in a specialized hierarchy
- The features can build relationships with each other

4. Application of the model for Virtual Commissioning

4.1 Assistance by means of mechatronic cell model in the Virtual Commissioning

As described above, the starting point to establish a workplace for Virtual Commissioning is the layout and the kinematic model of the plant. In addition, it can be assumed that the information about the resources and process is located in component related files (kinematics, geometry, behavioral descriptions), in databases or tools.

The goal is to build a workplace for Virtual Commissioning. A tool should be implemented to assist the configuration of the workspace. This assistance tool is called VIB configurator or VIBCon (Figure 10).

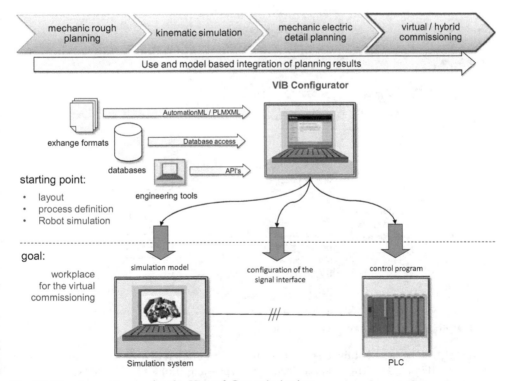

Fig. 10. The demonstrator for the Virtual Commissioning

The methodological foundations of the assistance tool allow the rule-based processing of raw data, which is related to a specific working step. The raw data can be extracted from various data sources. The result is the necessary configuration data or files for the workspace of Virtual Commissioning (Figure 11). In this way a semantic-based model is established in the mechatronic plant model. The model contains relationships such as the structure-based relationship "consistOf" or "isCharacterizedBy" (Figure 7). In addition, there are naming conventions that make, by using prefixes or suffixes, uniqueness semantic out of the mechatronic model. In this way, a check list based on the previously finished work can be generated. Based on this list, the library elements are instantiated. All these rules are defined in the VIB configurator. The following functions were implemented by the mechatronic plant model:

- Specification of the relationships between model elements and the workflow and the corresponding rules
- Generation of the needed configuration information by semantic inference or support of their derivation through a selection of possible alternatives

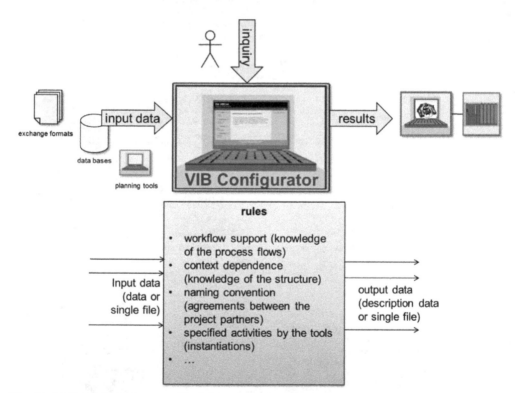

Fig. 11. Methodological approach of the rules

An example can be found at the beginning of the configuration of a Virtual Commissioning workstation (Figure 11). The needed document types (e.g. RobCAD file, VRML file, PLC program...) should be selected according to the context. These files will be earmarked and stored in the VIB configurator. When a new project is started, which uses the pattern of an old project, only the references to the types of these files and information are needed. These references are derived from the structural knowledge of the relationships between the types of resource and the types of their description.

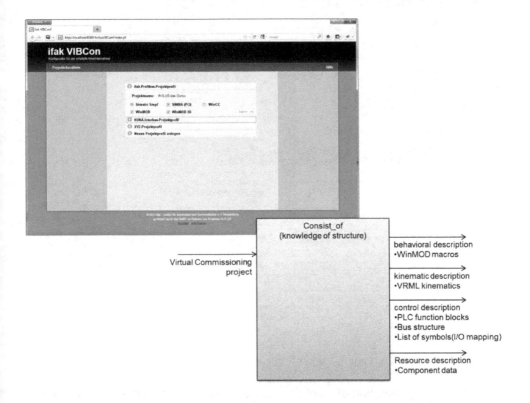

Fig. 12. Sample rule from the knowledge of structure – creating a project

A second example is the acquisition of information for all modules or components (implementation of the resources) that are included in the layout of a plant (Figure 13). The proper types of the files and information are needed for each resource in the project. The relationships between the model elements are integrated in the mechatronic plant model. Following the chain of these relationships between the elements leads to the references of the data sources. Naming conventions are applied here, such as extensions of the file name.

The mechatronic plant model serves as a knowledge model between the heterogeneous data sources and the data which is to be generated (Figure 13). The semantic model contains the know-how of the plant. The proper types of the files and information are needed for each resource in the project. The whole model is embedded in a software environment, which is described in subsection 4.2.

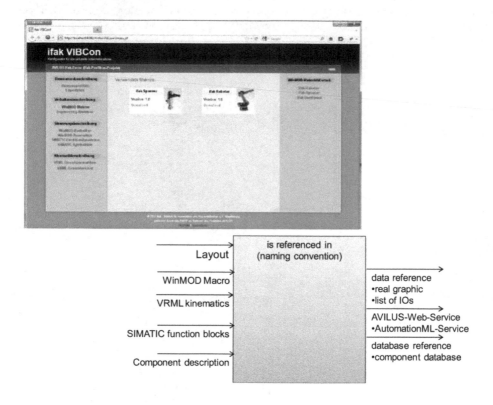

Fig. 13. Sample rule from naming convention – acquisition of information resource

4.2 Practical implementation of mechatronic cell model based on the semantic model

The main focus of the mechatronic model is the semantic modeling of the mechatronic components and the processes of the plant. This includes not only the description of the structural characteristics, e.g. the mechanical or electrical features, but also details like resource behaviors and process steps. Furthermore, a unified structure for the formal description of system properties is set (see Subsection 3.1). A framework (Figure 14) is implemented in order to access the information contained in the model like rules and relationships. Therefore, the needed information is found from different data resources (e.g. web services, exchange files of different formats). This process can be called reasoning.

The components and tasks of this framework include:

- Design and modeling (Figure 14, green)
- Standard technologies of "Semantic Web" (Figure 14, gray)
- Integration and evaluation (Figure14, orange)

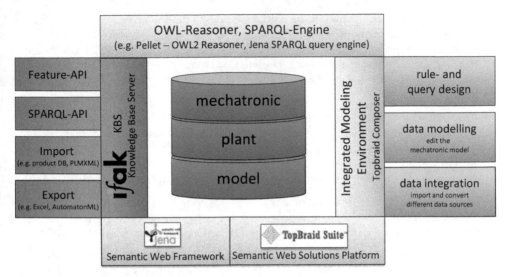

Fig. 14. Framework of the mechatronic plant model

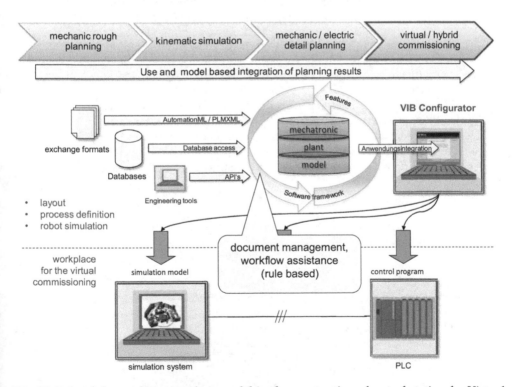

Fig. 15. Role of the mechatronic plant model in the preparation of a workstation for Virtual Commissioning

The most important technologies of the methods are data modeling and design of the rules and inquiries. For data modeling, as mentioned above, a semantic-based approach was deployed. The design of the rules and inquiries was based on this approach, which forms the basis of the knowledge processing known as automated rules-based reasoning. The instance data, which is based on the mechanisms of import and conversion, was integrated into the data model.

The integration of existing standard technologies became the focus in the development of the mechatronic plant model. The necessary technologies for knowledge processing have been implemented by applying the well-established tools and frameworks (e.g. OWL reasoned and SPARQL engines) for the Semantic Web. In this way, a semantic-based model with the functions, such as consistency check, reasoning and graph-based query, was built.

Different interfaces, which are based on a synchronous communication, are implemented as prototypes for the integration of the model into the engineering process. These interfaces enable the feature-based access (Feature API) and the query-based access (SPAQL API).

The concept, based on standardized exchange formats such as PLMXML and AutomationML (Draht et al., 2008), enables the flexible import and export of engineering data. The "Knowledge Base Server" (ifak-KBS) is a middleware which facilitates the integration of the mechatronic model using different interfaces into the engineering tools.

A conclusion of the important features of the VIB configurator is listed as follows:

- The development of a Virtual Commissioning workstation is a combination of tasks such as selection, classification and design, which are specified in a defined workflow. Assistance functions can be offered based on the knowledge of workflow and structure as well as naming conventions.
- The fundament of the assistance functions is the mechatronic plant model that was implemented in a software framework, based on semantic technologies.
- The introduced functions serve only as examples here, which can be supplemented by other detailed analysis of the process.
- Checklists and possible tests can be derived out of the available digital data.

5. Acknowledgment

This work is funded by the German Federal Ministry of Education and Research (BMBF) within the project AVILUS.

6. References

Bär, T. (2004). *Durchgängige Prozesskette vom digitalen Produkt bis zur realen Fabrik*, Congress „Digitale Fabrik" Ludwigsburg, Germany (2004)

Bergert, M.; Diedrich, C. (2008). Durchgängige Verhaltensmodellierung von Betriebsmitteln zur Erzeugung digitaler Simulationsmodelle von Fertigungssystemen, *atp – Automatisierungstechnische Praxis*, Vol.7 (2008), pp. 61-66, Oldenbourg Industrieverlag, ISSN 2190-4111

Bracht, U. (2002). Ansätze und Methoden der Digitalen Fabrik, *Tagungsband Simulation und Visualisierung*, Magdeburg, pp. 1-11

Draht, R.; Lüder, A.; Peschke, Jörn. & Hundt, L. (2008). AutomationML - The glue for seamless automation engineering, *Proceedings of the 13th international conference on Emerging Technologies and Factory Automation*, pp.616-623, ISBN 978-1-4244-1505-2, Patras, Hamburg, September 15-18, 2008

Heeg, M. (2005). *Ein Beitrag zur Modellierung von Merkmalen im Umfeld der Prozessleittechnik*, VDI-Verlag, ISBN 978-3-18-506008-3, Düsseldorf, Germany

Jain, Atul; Vera, D. A.; Harrison, R. (2010). Virtual commissioning of modular automation systems, *Proceedings of the 10th IFAC Workshop on Intelligent Manufacturing Systems*, pp. 79-84, ISBN: 978-3-902661-77-7, Lissabon, Portugal, July 1-2, 2010

Kiefer, J. (2007). *Mechatronikorientierte Planung automatisierter Fertigungszellen im Bereich Karosserierohbau*, PhD thesis, Universität des Saarlandes, Schriftenreihe Produktionstechnik, ISBN 978-3-930429-72-1, Saarbrücken, Germany

Kiefer, J.; Bergert M. & Rossdeutscher M. (2010). Mechatronic Objects in Production Engineering, *atp – Automatisierungstechnische Praxis*, Vol.12 (2010), pp. 36-45, Oldenbourg Industrieverlag, ISSN 2190-4111

McBride, B. (2010). An Introduction to RDF and the Jena RDF API, available from http://jena.sourceforge.net/tutorial/RDF_API/index.html, Retrieved on 22/02/2012

McGuinness, D.; van Harmelen, F. (2011). OWL web ontology language overview, available from http://www.w3.org/TR/owl-features/

Mühlhause, M.; Suchold, N. & Diedrich, C. (2010). Application of semantic technologies in engineering processes for manufacturing systems, *Proceedings of the 10th IFAC Workshop on Intelligent Manufacturing Systems*, pp. 61-66, ISBN: 978-3-902661-77-7, Lissabon, Portugal, July 1-2, 2010

Prud'hommeaux, E.; Seaborne, A. (2008). *SPARQL Query Language for RDF*, W3C Recommendation

Rauprich G.; Haus C. & Ahrens W. (2002). PLT-CAE-Integration in gewerkeübergreifendes Engineering und Plant-Maintenance, *atp - Automatisierungstechnische Praxis*, Vol.2 (2002), pp. 50-62, Oldenbourg Industrieverlag, ISSN 2190-4111

Schreiber, S.; Schmidberger, T.; Fay, A.; May, J.; Drewes, J. & Schnieder, E. (2007). UML-based safety analysis of distributed automation systems, *Proceedings of the 12th international conference on Emerging Technologies and Factory Automation*, pp.1069-1075, ISBN 1-4244-0826-1, Patras, Greece, September 25-28, 2007

TopQuadrant (2011). *Getting Started with SPARQL Rules (SPIN)*, Version 1.2, Available from http://www.topquadrant.com/spin/tutorial/SPARQLRulesTutorial.pdf, Retrieved on 22/02/2012

VDI 4499 Blatt 1 (2008). Digitale Fabrik – Grundlagen. VDI-Handbuch Materialfluss und Fördertechnik, Bd. 8, Gründruck. Berlin: Beuth

Volz, R. (2008). *Web Ontology Reasoning with Logic Databases*, PhD thesis, 17 Februar 2008, Universität Karlsruhe, Germany

Wollschläger M.; Braune A.; Runde S.; Topp U.; Mühlhause M.; Drumm O.; Thomalla C.; Saboc A. & Lindemann L. (2009), Semantische Integration im Lebenszyklus der Automation, Tagungsband zum Automation 2009 –Kongress Baden-Baden, VDI-Berichte 2067, pp. 167-170, VDI-Verlag, Düsseldorf, Germany

Wünsch, G. (2008). *Methoden für die Virtuelle Inbetriebnahme automatisierter Produktionssysteme*, PhD thesis, Technische Universität München, Herbert Utz Verlag, ISBN 978-3-8316-0795-2, München, Germany

Zäh, M. F.; Wünsch, G. (2005). Schnelle Inbetriebnahme von Produktionssystemen, *wt Werkstattstechnik 95*, Vol. 9 (2005), pp. 699-704

7

Power System and Substation Automation

Edward Chikuni

Cape Peninsula University of Technology
South Africa

1. Introduction

Automation is "the application of machines to tasks once performed by human beings, or increasingly, to tasks that would otherwise be impossible", Encyclopaedia Britannica [1]. The term automation itself was coined in the 1940s at the Ford Motor Company. The idea of automating processes and systems started many years earlier than this as part of the agricultural and industrial revolutions of the late 18th and early 19th centuries. There is little disputing that England was a major contributor to the Industrial Revolution and indeed was the birth place of some prominent inventors, for example. in the area of textiles:

- James Hargreaves: Spinning Jenny
- Sir Richard Arkwright: Mechanical Spinning Machine
- Edmund Cartwright: Power Loom

Cartwright's power loom was powered by a steam engine. At these early stages we see the symbiotic relationships between automation, energy and power. The early forms of automation can only largely be described as mechanisation, but the emergence of electrical power systems in the late 19th century and the entry of electronic valves in the early 20th century heralded the humble beginnings of modern automation. With electronic valves came computers. One of the earliest computers was the ENIAC (Electronic Numerical Integrator and Automatic Computer) built over two years between 1943 and 1946. It occupied an area of 1000 square feet (about 93 square metres), had 18000 valves and consumed 230 kW [2].

Before the deployment of computers in industrial automation, relays and RELAY LOGIC, the wiring of circuits with relays to achieve automation tasks, was in common use. Today, however, relay logic is far less used than computer-based, PROGRAMMABLE LOGIC, which has followed the invention of the transistor, integrated circuits and microprocessors.

2. Automation in the automobile (car / truck) industry

The motor assembly line pioneered by Ransom Olds and Henry Ford, the maturity of computer technology and the structured nature of the car assembly process led early entrepreneurs in the motor industry to view automation as key to business success. Indeed there was all to be gained in automation and today automation is viewed to have, among many other attributes, the following [3]:

a. relieves humans from monotony and drudgery
b. relieves humans from arduous and dangerous tasks
c. increases productivity and speeds up work rates
d. improves product quality
e. reduces costs and prices
f. increases energy and material savings
g. improves safety
h. provides better data capture and product tracking.

The automotive industry was very competitive right from the early days and automation soon came to be seen as key to commercial success. General Motors embraced it and so did many other auto manufacturing corporations in the US, Europe and Japan. The computer system used was designed for the usually harsh industrial environments. Programmable Logic Controllers (PLC) as such computers are known typically to have many inputs and outputs; the inputs receiving sensor signals, the outputs being for displaying information or driving actuators (Figure 1).

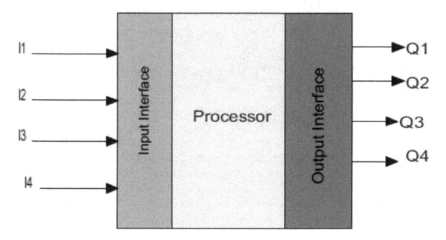

Fig. 1.

The need for standardization was realized early, especially when many PLCs are required to achieve the automation task. The standards generating body IEEE has been a driving force in computer communication standards over the years. One of its standards, the token ring 802.4 was implemented in modified form by General Motors in its Manufacturing Automation Protocol (MAP).

3. Power system automation

The early power plants had a modest number of sensor and action variables, of the order of several hundreds. Modern large power stations have in excess of hundreds of thousands, even tens of million variables [3]. It is therefore easy to see that automation took root in power and generating stations earlier than in transmission and distribution. One of the useful applications of automation is in the railway industry where remote control of power is often vital. This manifests itself in systems to control and manage power supplies to electric locomotives. The deprivation of power to any locomotives in a section could seriously disrupt schedules, inconvenience customers and in the end have serious adverse financial implications. Consider Figure 2

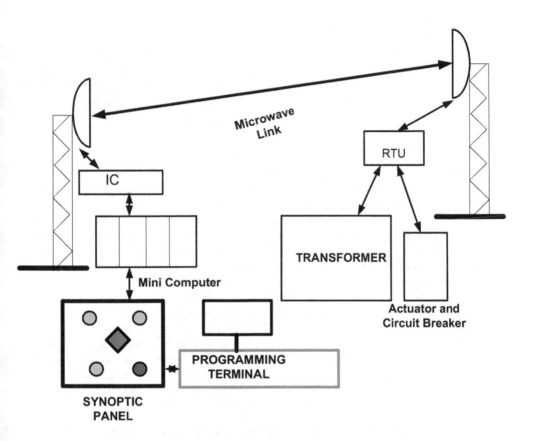

Fig. 2.

In earlier systems, the "mini computer" would have been sourced from specialised computer companies such as IBM, Perkin Elmer, Digital Equipment Corporation. The operating systems would have belonged to the same equipment provider (e.g., VMS VAX, in the case of Digital Equipment Corporation). The microwave link would have been part of the infrastructure costs for the project. The synoptic panel, allowed visual representation of the system states to the operator (and also some capabilities for remote switching). For Figure 1, the operator at the control centre is able switch a circuit breaker ON or OFF through the microwave link. Interface cards (IC) provide the necessary I/O capabilities. The remote terminal unit (RTU) is able to send a switching signal to the circuit breaker through a suitably designed actuator. The circuit breaker is also able to send its position (ON or OFF) through the RTU and the same microwave link to the operator's panel. The older type of hardwired panel is largely being replaced by electronic displays and the controller is able to receive and view system status through touch panel capabilities or through networked computers in the control room.

The proprietary nature of the old "legacy systems meant that embarking on the path to automation was not a trivial matter. The cost of ownership, (infrastructure, hardware and software costs) was very high. Also very high were the costs of maintaining and upgrading this hardware and software. The cost of software licences was often prohibitive. On the other hand cyber threats and virus attacks were unheard of. The systems themselves were broadly secure, but not necessarily reliable. Communication link (microwave failure) meant that there was at the time, as there is even in this generation, a need for "manual back-ups". This usually meant sending a technician to do manual switching operations.

4. Modern grid and substation automation

Power system automation happens in segments of the power system (Northcote-Green, Wilson) [4] which can serve different functions. One segment is bulk transmission of power which traditionally was handled by the power producer, but increasingly (in de-regulated environments), is handled by an independent transmission system operator (TSO). Bulk transmission is usually associated with outdoor switchyards and high voltage operating voltage levels (in excess of 132 kV). Bulk transmission substations play a critical role in energy trading and power exchanges. Wholesale electricity is sold through the transmission system to distributors. Figure 3 shows a portion of the 400 kV outdoor substation at AUAS near Windhoek, the capital of Namibia. Namibia is a net importer of electrical power most of it from neighbouring South Africa and this substation is of vital importance. The substation with which it connects in South Africa is over 800 km away at Aries near Kenhardt. Figure 4 shows the Namibian electrical power transmission network. The complex interconnections between equipment, such as transformers, reactors, lines and bus-bars, is such that manual operation is not a practical proposition. In the case of the AUAS substation, effective control is in the hands of Namibia Power Corporation's (Nampower) headquarter-based National Control Centre in Windhoek.

The other segment of automation is at distribution level. Large distributors are typically municipal undertakings or in countries in which electrical power is de-regulated, the so called "DISCOS". Automation has existed at the distribution level for many years, but has been restricted to situations involving either large numbers of customers or critical loads. As a result of this the quality of service given has been very good, while the rural consumers have been at a disadvantage. For power distributors, automating rural supplies was not cost-effective due to the dispersed nature of the lines and loads (Alsthom Network Protection and Application Guide) [5]. Technology changes in recent years, national power quality directives as well as increased consciousness by consumers themselves has led to radical changes in our Power System Control infrastructure and ways of operation with serious implications for those organisations that lag behind [NPAG].

Fig. 3. A group of Polytechnic of Namibia students visit the AUAS 400 kV substation

4.1 Distribution systems automation

From experience, faults at transmission levels are less frequent than at distribution levels [6]. At the same time distribution networks are not only complex, but the consequences of failure are quite severe. For this reason investment in distribution automation will increase. The elements that characterise distribution automation systems are given the definition by the IEEE. According to the IEEE, a Distribution Automation System (DAS) is "a system that enables an electric utility to remotely monitor, coordinate and operate distribution components, in a real-time mode from remote locations [7]". In this chapter we shall deal in more detail with the components involved and how they are coordinated within a DAS. In countries or situations where there are large networks, the network (primary distribution) itself is subdivided into more segments, namely, one for large consumers (no transformation provided) and for the rest at a lower HV voltage (secondary distribution) (Figure 6).

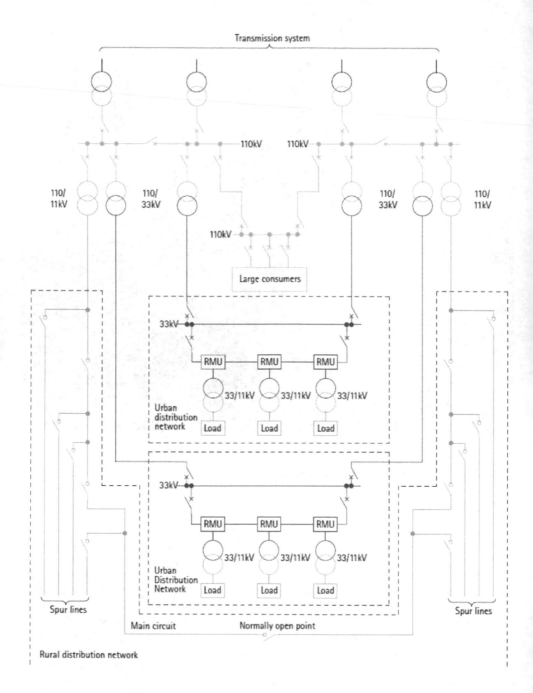

Fig. 4. (courtesy AREVA, NPAG)

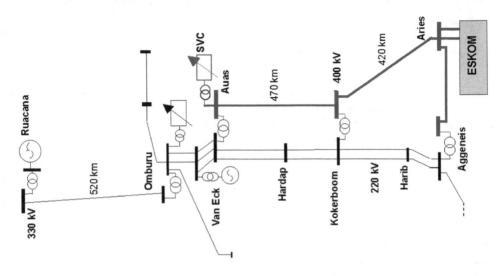

Fig. 5. (courtesy, Namibian Power Corporation)

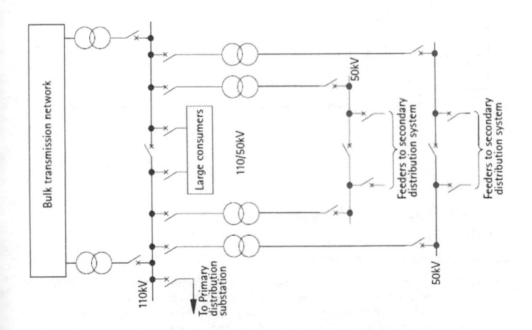

Fig. 6. (courtesy AREVA, NPAG)

5. Power system automation components

Power system automation components may be classified according to their function:

- Sensors
- Interface Equipment
- Controllers
- Actuators

Thus we see that Figure 1 is still a good representation of what is needed to effect automation, whether it is for EHV transmission, sub-transmission or distribution. Figure 7 depicts the control philosophy of a power system automation scheme.

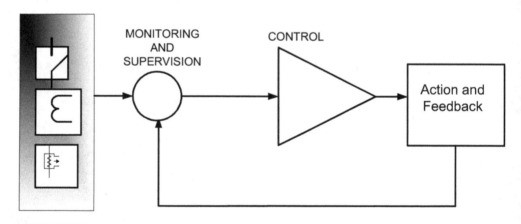

Fig. 7.

5.1 Overview of power system components

5.1.1 Sensors

5.1.1.1 Current and voltage transformers

Individually or in combination current and voltage transformers (also called instrument transformers) are used in protective schemes such as overcurrent, distance and carrier protection. Also in combination current and voltage transformers are also used for power measurements. In general custom specified voltage and current transformers are used for power metering, because of the increased accuracy requirements. Figure 8 shows instrument transformers in one of the substation areas (called bays).

Fig. 8.

5.1.1.2 Other sensors

For reliable electrical power system performance the states, stress conditions and the environmental conditions associated with the components have to be monitored. A very costly component in a substation is a transformer. For a transformer, monitoring is done, for example, for pressure inside the tank, winding temperature and oil level. For circuit breakers, sensing signals may need to be obtained from it such as gas pressure and number of operations.

5.1.2 Switches, isolators, circuit breakers

A most important function of a substation is the enabling of circuit configuration changes occasioned by, for example, planned maintenance, faults feeders or other electrical equipment. This function is of course in addition to the other important function of circuit protection which may also necessitate configuration changes. Modern switches and circuit breakers will have contacts or sensors to indicate their state or position. Figure 9 shows the ABB HH circuit breaker mechanism. The plant required to achieve the desired operation is usually quite elaborate and includes controls and protection to ensure that it operates reliably.

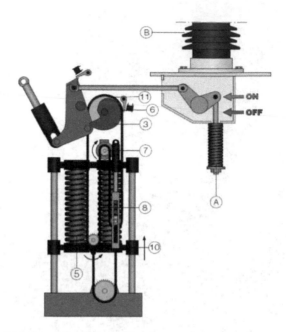

Fig. 9. Portion of HH ABB circuit breaker mechanism

6. IEC 61850 substation automation: Origin and philosophy

The International Electrotechnical Commission is one of the most recognisable standard generating bodies for the electrical power industry. Its standard the IEC 61850 "Communication Networks and Systems in Substations" is a global standard governing communications in substations. The scope of the standards is very broad and its ramifications very profound. So profound in fact that it is hard to imagine any new modern substation that would not at least incorporate parts of this standard. In addition, the standard is almost sure to be adopted albeit in customised / modified form in Generation, Distributed Energy Resources (DER) and in manufacturing. The standard has its origins in the Utility Communications Architecture (UCA), a 1988 initiative by the Electrical Power Research Institute (EPRI) and IEEE with the initial aim of achieving inter-operability between control centres and between substations and control centres. In the end it was found to be more prudent to join efforts with similar work being done by the Working Group 10 of Number 57 (TC57). The emerged document IEC 61850 used work already done by the UCA as a basis for further development.

6.1 IEC 61850 substation architecture

6.1.1 Substation bays

In an IEC 61850 compliant substation, equipment is organized into areas or zones called bays. In these areas we find switching devices (e.g., isolators and circuit breakers) that connect, for example, lines or transformers to bus-bars.

Examples of the bays would be:

- Incomer bay
- Bus-coupler bay
- Transformer bay

Figure 8 above, for example, could depict a transformer bay.

6.1.2 Merging units

Merging units are signal conditioners and processors. For example, they accept, merge and synchronise sampled current and voltage signals (all three phases' quantities of the CT/VT) from current and voltage transformers (conventional and non-conventional) and then transfer them to intelligent electronic devices (see IEDs, in the next section). So called electronic VTs and CTs are being manufactured by some companies which use new ways of sensing with the overall size being reduced. With electronic sensing, the sensing and merging are combined. Figure 10 gives an overview of the functions and associated inputs of a merging unit.

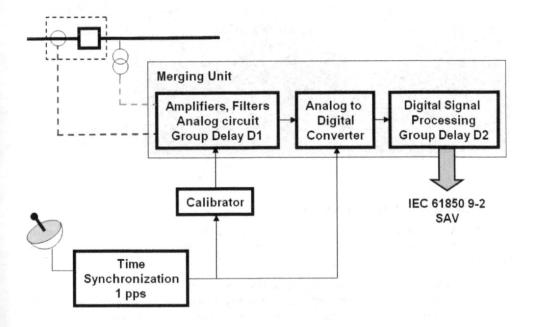

Fig. 10. (Jansen & Apostolov)

As technology progresses it is believed that there will be a move away from copper connections from field devices to the substation control room in favour of fibre. Figure 11 shows a merging unit (Brick) by vendor GE.

6.1.3 Intelligent Electronic Devices (IEDs)

An IED is any substation device which has a communications port to electronically transfer analog, status or control data via a proprietary or standard transmission format (BPL Global IEC 61850 Guide) [8]. Examples of IEDs are:

- Modern IEC 61850 protection relays (distance, over-current, etc.)
- Equipment-specific IED (e.g., for transformer bay protection and control, with tripping logic, disturbance monitoring, voltage, current, real and reactive power, energy, frequency, etc.).
- Bay controllers

Figure 11 shows some IEDs from various vendors with multiple functionality.

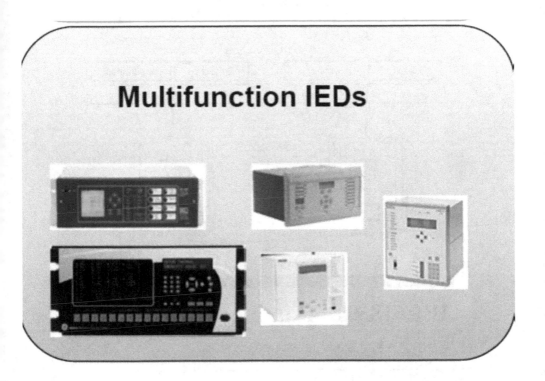

Fig. 11.

In reality today's IEDs have "mutated" to the form of programmable logical controllers (PLCs) of another kind with multiple capabilities.

6.1.4 Device/system integration: Substation functional hierarchy

An IEC 61850-designed substation has the following hierarchical zones:

- Process
- Bay
- Station

Diagrammatically this is illustrated in Figure 12 (Jansen & Apostolov) [9]. A complete representation that includes aspects, such as links to remote control centres and GIS, is given in Figure 13.

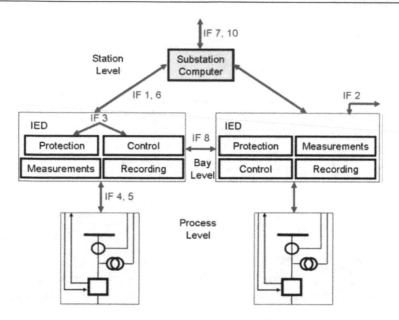

Fig. 12.

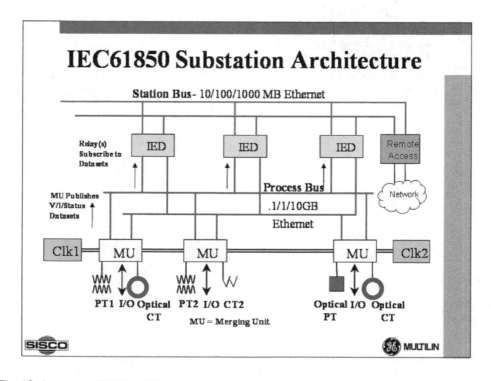

Fig. 13. (courtesy SISCO & GE)

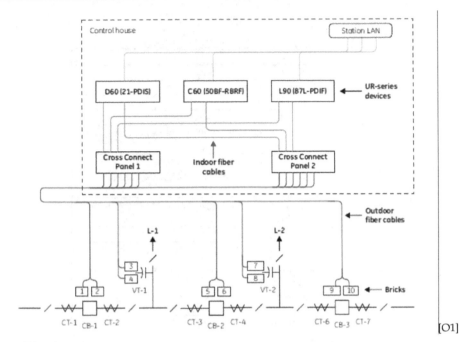

Fig. 14. Fibre-based

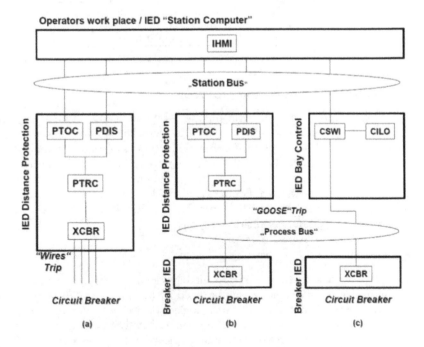

Fig. 15.

STRUCTURE OF THE IEC 61850 STANDARD

Part #	Title
1	Introduction and Overview
2	Glossary of terms
3	General Requirements
4	System and Project Management
5	Communication Requirements for Functions and Device Models
6	Configuration Description Language for Communication in Electrical Substations Related to IEDs
7	Basic Communication Structure for Substation and Feeder Equipment
7.1	- Principles and Models
7.2	- Abstract Communication Service Interface (ACSI)
7.3	- Common Data Classes (CDC)
7.4	- Compatible logical node classes and data classes
8	Specific Communication Service Mapping (SCSM)
8.1	- Mappings to MMS(ISO/IEC 9506 – Part 1 and Part 2) and to ISO/IEC 8802-3
9	Specific Communication Service Mapping (SCSM)
9.1	- Sampled Values over Serial Unidirectional Multidrop Point-to-Point Link
9.2	- Sampled Values over ISO/IEC 8802-3
10	Conformance Testing

7. Substation communications and protocols

With the IEC 61850 technology and with all the components and systems described in previous sections functioning normally, we have in fact a virtual substation. The remote terminal units (RTU) increasingly with IED functionality, pass on analog and digital data through either copper or fibre to IEDs in the substation control room in the form of relays or bay controllers. The process of transferring data and communicating it to various devices has been greatly simplified with the aid of the standard. The data arriving at the IEDs comes already formatted / standardized. The situation is similar to the "plug and play" philosophy applied to computer peripherals of today.

7.1 Virtualisation

With the IEC 61850 a real substation is transformed into a virtual substation, i.e., real devices transformed into objects with unique standardized codes. In Figure 16, a real device, a transformer bay is transformed into a virtual, logical device with descriptive name, e.g., Relay1. Inside the device are logical nodes (LN) named strictly in accordance with the IEC standard. For example, a circuit breaker inside this logical device is given XCBR1 [10]. In turn the breaker has other objects associated with it, e.g., status (open / closed) and health. The services associated with this data model are defined in the Abstract Communications System Interface (ACSI). The following ACSI functions are listed by Karlheinz Schwartz [11]:

- Logical Nodes are used as containers of any information (data objects) to be monitored
- Data objects are used to designate useful information to be monitored
- Retrieval (polling) of the values of data objects (GetDataObjectValues)
- Send events from a server device to a client (spontaneous reporting)
- Store historical values of data objects (logging)
- Exchange sampled values (current, voltages and vibration values)
- Exchange simple status information (GOOSE)
- Recording functions with COMTRADE files as output

7.2 Mapping

IEC 61850 is a communications standard, a main aim of which is interoperability. A good definition is:

"Interoperability is the ability of two or more IEDs (Intelligent Electronic Devices) from the same vendor, or different vendors to exchange information and uses that information for correct co-operation" [12]. Although ACSI models enable all IEDs to behave identically from a general network behaviour perspective, they still need to be made to work with practical networks in the power industry, (Baigent, Adamiak and Mackiewicz) [10]. This universal compatibility is achieved through mapping of the abstract services to universal, industry-recognised protocols. Presently the protocol most supported is the Manufacturing Message Specification (MMS). MMS was chosen because it has an established track record in industrial automation and can support the complex and service models of IEC 61850.

Table 1 gives an idea of the naming process:

IEC61850 TO MMS OBJECT MAPPING

IEC 61850 Objects	MMS Object
SERVER class	Virtual Manufacturing Device (VMD)
LOGICAL DEVICE class	Domain
LOGICAL NODE class	Named Variable
DATA class	Named Variable
DATA-SET class	Named Variable List
SETTING-GROUP-CONTROL-BLOCK class	Named Variable
REPORT-CONTROL-BLOCK class	Named Variable
LOG class	Journal
LOG-CONTROL-BLOCK class	Named Variable
GOOSE-CONTROL-BLOCK class	Named Variable
GSSE-CONTROL-BLOCK class	Named Variable
CONTROL class	Named Variable
Files	Files

IEC61850 SERVICES MAPPING (PARTIAL)

IEC61850 Services	MMS Services
LogicalDeviceDirectory	GetNameList
GetAllDataValues	Read
GetDataValues	Read
SetDataValues	Write
GetDataDirectory	GetNameList
GetDataDefinition	GetVariableAccessAttributes
GetDataSetValues	Read
SetDataSetValues	Write
CreateDataSet	CreateNamedVariableList
DeleteDataSet	DeleteNamedVariableList
GetDataSetDirectory	GetNameList
Report (Buffered and Unbuffered)	InformationReport
GetBRCBValues/GetURCBValues	Read
SetBRCBValues/SetURCBValues	Write
GetLCBValues	Read
SetLCBValues	Write
QueryLogByTime	ReadJournal
QueryLogAfter	ReadJournal
GetLogStatusValues	GetJournalStatus
Select	Read/Write
SelectWithValue	Read/Write
Cancel	Write
Operate	Write
Command-Termination	Write

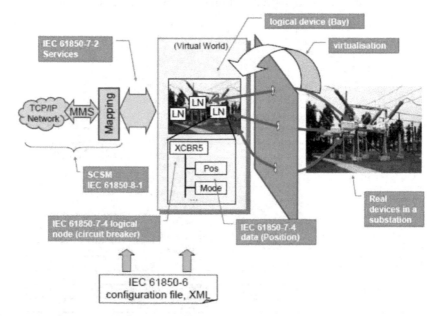

Fig. 16. Karlheinz Schwartz

8. Communication of events in an IEC 61850 substation

In his IEC 61850 Primer, Herrera states that "IEC 61850 provides a standardized framework for substation integration that specifies the communications requirements, the functional characteristics, the structure of data in devices, the naming conventions for the data, how applications interact and control the devices, and how conformity to the standard should be tested. In simpler terms, IEC 61850 it is an open standard protocol created to facilitate communications in electric substations."

8.1 The communication structure of the substation

The IEC 61850 architecture there are two busses:

- Process bus
- Station bus

IEC 61850 **station bus** interconnects all bays with the station supervisory level and carries control information such as measurement, interlocking and operations [13].

IEC 61850 **process bus** interconnects the IEDs within a bay that carries real-time measurements for protection called sampled values or sampled measured values [13].

Figure 17 shows the basic architecture.

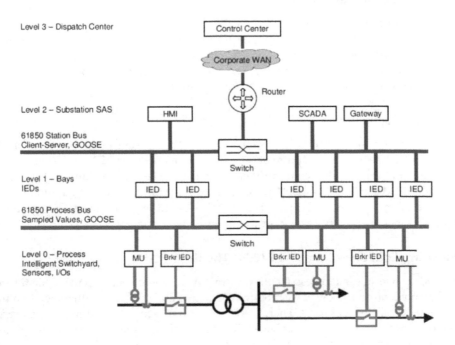

Fig. 17.

The process bus is designed to be fast since it must carry crucial I/O between IEDs and sensors/actuators.

The requirements for the process bus cited in various literature sources are as follows:

- High environmental requirements for the terminal equipment (electromagnetic compatibility, temperature, shock, where applicable) in the area of the primary system
- Adequate bandwidth for several SV data streams
- Highly prioritized trip signals for transmitting from the protection device to the CBC
- Permeability of data to the station bus/data filtering at the coupling point
- Simultaneous TCP/IP traffic for normal control and status signal traffic as well as reports on the process bus
- Download/upload channel for setting or parameterizing functions
- Highly precise time synchronization
- Redundancy
- For reasons of speed, the process bus is based on optical fibre with high data throughput of about 10Gbits/s. Because of its enhanced data capacity it is capable of carrying both GOOSE (Generic Object Oriented Substation Event) and SMV (Sampled Measured Values). The station bus is used for inter-IED communications. Only GOOSE messaging occurs in the station bus.

9. Substation control and configuration

Although the strengths of the IEC 61850 in the capturing, virtualisation, mapping and communication of substation information are undoubted, it will still be necessary to link everything together and to design a control strategy. This strategy must utilize the experience and expertise of the asset owner. The substation must also respond in accordance with the operational and safety criteria set by the organization.

9.1 Substation configuration

Automation of the substation will require in the first instance the capture of its configuration. This requires the capture of the information on all the IEDs in the substation. In some cases the IEDs could be from different vendors. The information has to be in a standardized IED Capability Description (ICD). Then, using a system configuration tool, a substation description file is created (Figure 18). The SCD (Substation Configuration Description) is then used by relay vendors to configure individual relays [14].

10. Wider implications of the IEC 61850: The Smart Grid

"Smart Grid" is a term used to describe the information driven power systems of the future. This will involve introducing new electronic, information and computer technology into the whole value chain of electrical energy systems from generation, transmission and distribution down to the consumer level. Figure 19 shows the linkages between the technology of electricity production and commerce.

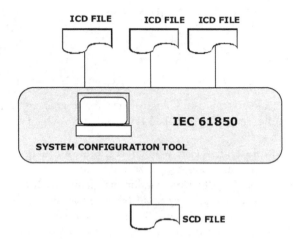

Fig. 18.

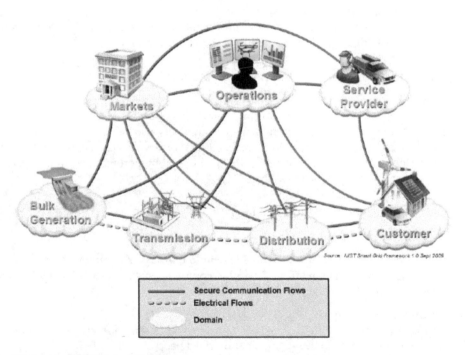

Fig. 19. Source NIST Smart Grid Framework

We have seen that automation started on the factory floor and some of the IEC 61850 functions use manufacturing protocols such as MMS. We already start to see the trend towards extending the IEC 61850 to generating stations. It is therefore not hard to imagine IEC 61850 like protocols encompassing every facet of engineering, including manufacture.

10.1 Smart Grid benefits

Among the benefits of Smart Grid are:

- Increased grid efficiency - the use of control systems to achieve optimum power flow through, for example, centrally controlled FACTS devices which can increase efficiency of the transmission system
- Better demand control - a Smart Grid would incorporate an energy management system to manage demand (e.g., managing the peaks and valleys)
- Asset optimization - the IEC 61850 information model already has the capability to store not only the status of a logical device / node, but also condition / health
- Management of renewable energy sources - renewable energy sources, such as wind and solar, tend to be unpredictable, therefore, the Smart Grid system can enable predictions on the availability of these resources at any moment and ensure proper energy scheduling decisions are taken
- Management of plug in electric vehicles - the Smart Grid can inform electric vehicle motorists of the nearest charging stations
- Smart metering - with smart metering, power usage and tariffs can be administered remotely to the advantage of both the supplier and the consumer

11. Security threats in automated power systems

In this chapter we have seen the central role computer hardware and software in the control and management of the power system bring tremendous benefits. However, investing in these high technology, information technology reliant assets also brings threats. The threats are quite serious especially when it is realized that every critical component of the substation becomes a virtual computer. The IED mentioned numerous times in this chapter is itself a computer. What are the threats?

11.1 SCADA vulnerabilities Chikuni, Dondo [15]

- Computing vulnerabilities

Hardware: RTUs, IEDs and SCADA Masters belong to the class of computer hardware and suffer from the same vulnerabilities of regular computer systems such as interruption (denial of services [DoS]) and eavesdropping

Communication links: the vulnerabilities are also similar to those in regular computer networks - if messages are not encrypted, data or passwords can be intercepted. Radiation emissions from equipment can be read by unauthorized people

- **SCADA software:** the most common attacks come in the form of interruption, interception and modification. Software bugs, if not fixed in time, can attract hobbyist hackers to attack unpatched SCADA [15]

- **Data:** SCADA data has more value to the attacker than hardware and software. Data may be stolen by competitors of saboteurs. To safe guard the data, encryption needs to be included

11.2 Other vulnerabilities

- **Equipment location:** we have seen that some IEDs and RTUs are located in usually unmanned remote locations; where this applies these must be housed or mounted securely
- **Remote access:** access to relays, controllers, IEDs and RTUs should be password protected; encryption modems are available for secure dial-up communications
- **Human element:** the employee could be the most vulnerable part of the automated power system, therefore, no unauthorized persons should have access to the control terminals. Strong authentication and smart card access are recommended
- **Integrity and confidentiality:** software and hardware should have at least the US National Computer Security Centre (NCSC) class 2 rating. In Europe criteria similar to that of NCSC is managed through the European Information Technology Security Evaluation Criteria (ITSEC)

11.3 Attack examples

Nadel et al. [16] list what they describe as generic attack categories to which all network-based threats to substation automation systems can be reduced, namely:

- Message modification
- Message injection
- Message suppression

They demonstrate the various paths that an attacker can take in a given attack. An example of circuit breaker attack scenario is shown in Figure 20. In this case an attacker may take one of the paths in the graph to prevent a circuit breaker from opening when it is supposed to.

11.4 Countermeasures

Nadel et al. [16] list some of the important assumptions / precautions necessary before any meaningful countermeasures can be instituted. These include static configuration of the SAS and the number and types of devices in the bay level are known; configuration changes only to occur during major maintenance or modification work; also that the SAS is not used for billing and no general purpose PCs are allowed at bay level.

Message modification

In this attack parts of a valid, existing message are modified in transit. Detection is facilitated through encryption and digital signatures, with the receiver having a record of all authorized senders.

Message injection and replay

The attacker sends messages which are not intended to be sent by any authorized sender. These may be entirely new messages from the attacker or original untampered with replayed messages. Digital signatures are a way of combating message injection. To mitigate against replay, a digital signature and message sequence number are required.

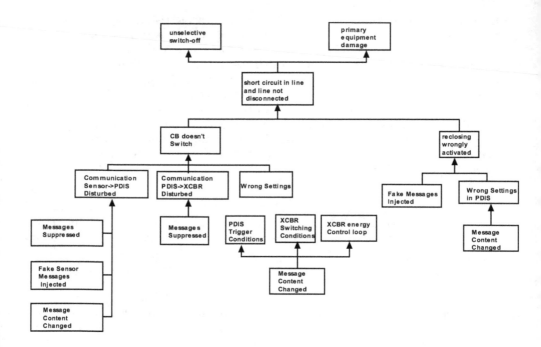

Fig. 20. Example of attack graph for a circuit breaker

Message suppression

In this attack certain messages between SAS devices are prevented from reaching the receiver, e.g., circuit breaker control devices are isolated from protection devices. Message suppression can involve several other types of attack, e.g., re-configuration of routers or switches, cutting wires or congesting the network so that genuine messages cannot get through (denial of service attack).

Security protocols

The multiplicity and varied nature of SAS attacks makes it imperative to institute robust security protocols capable of handling all eventualities. Such protocols include the use of private keys (only known to the sender), encryption and sequence numbers (initial number known between sender and receiver at the start).

Markets	
Subsystem Communication	
Intra-Markets	See Clause Markets
Intersystem Communication	
Operation	For scheduling and trading purposes, information about the availability of power (transfer power, operating reserve) or order information is transmitted to or from the operation system
Bulk Generation	For scheduling and trading purposes, information about the availability of power (transfer power, operating reserve) is transmitted from the bulk generation system
Service	Support functions for markets (e.g. forecasting for renewable generation)
Prosumer	For scheduling and trading purposes information about the availability of power or order information is transmitted to or from the markets system

System: Service

The service system offers potential for a wide range of new service developments. New business models may emerge due to the opportunities of the future Smart Grid. Therefore the service system will have and depend on various interfaces to other systems.

Service	
Subsystem Communication	
The new service application shall follow a standardized way of software development in order to seamlessly fit into an overall system. The relevant standards are not within the scope of the IEC	
Intersystem Communication	
Operation	Support functions for operation (e.g. forecasting for renewable generation)
Market	Support functions for markets (e.g. forecasting for renewable generation)
Prosumers	Customer services (Installation, Maintenance, Billing, Home & Building Management) are quite conceivable

System: Prosumer
Description

Prosumer	
Subsystem Communication	
See Clause HBES/BACS	
Process Automation	In many industries (e.g. chemical, manufacturing) process automation is applied to control and supervise not only the manufacturing process but also the energy consumption or generation
Intersystem Communication	
Service	Support functions for operation (e.g. forecasting for renewable generation)
Operation	See Clause AMI, DER
Markets	For scheduling and trading purposes information about the availability of power or order information is transmitted to or from the markets system
Distribution	Typically the distribution system infrastructure is used for the communication to DMS

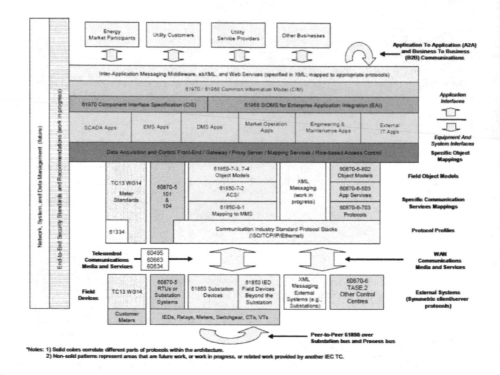

*Notes: 1) Solid colors correlate different parts of protocols within the architecture.
2) Non-solid patterns represent areas that are future work, or work in progress, or related work provided by another IEC TC.

12. Effects on educational curricula

To give an idea of the profound changes the power system will have on our education systems and also to give some suggestions on how to mitigate some of the challenges, we reproduce the following excerpts from a paper by Chikuni, Engelbrecht and Dongo [17] at the PowerCon 2010 conference:

"When we analyse a modern substation incorporating the new substation technology based on IEC 61850, it would seem that the role of an electrical engineer is notable by its absence. Certainly some of the responsibilities of both engineers and technicians have shifted and there needs to be new breed of electrical engineers altogether. Some questions need to be answered:

- Should we retrain the power engineer in networks or
- Should we train network engineers so that they acquire power engineering knowledge or
- Should we work with a completely new curriculum which merges power systems, electronics and networks into one programme?

We first need to acknowledge that there are already a lot of good power engineers out there trained in the traditional manner, i.e., starting off with physics, circuits and systems, electrical machines and power systems (including electrical protection). For these engineers, one needs to identify those who can benefit both themselves and their organizations by going through this additional training. The process of training engineers has been quite formal, especially if they wish to attain professional status. One typically needs four to five years of formal training and a further two years of guided industrial training before attaining the status of chartered (CEng) or professional (PrEng) engineer. A great debate will ensue, therefore, when a computer network engineer is designated the 'responsible person' in an electrical substation, notwithstanding the obviously immense power this individual will have in making sure that the substation operates correctly, safely and efficiently.

The other route is to include networking as part of any electrical engineering curriculum. A few programmes today include industrial automation and a few even include computer networking. In the University of Zimbabwe model all electrical engineering students have a chance to complete at least the first semesters of a CISCO network academy programme. Indeed some complete the CCNA (four semesters). Whatever solution is arrived at, it is clear that electrical engineering training curricula inevitably have to include more and more electronics, sensors, automation and networking, not as peripheral subjects, but as part of the core.

13. Summary and conclusions

In this chapter we have seen the extremely rapid development of automation, starting from the years of mechanisation, production lines and the taking root of computer-based automation in the car manufacturing industry. Then we noticed rapid increases in computer power in both hardware and software forms. There has also been tremendous moves in standardization in North America and Europe. We have seen too IEC 61850 international cooperation in standards development and the benefits that are already being reaped from this. Interoperability brings some relief to customers, giving them the ability to choose hardware from an increasing variety of vendors. Quite striking is the increasing dominance of ICT in power system control and massive changes in power system operation and practice. The power systems have become more complex - more interlinked. The complexity presents new challenges. The traditionally trained power systems engineer lacks the know how to understand or tackle faults that could arise in these systems. On the other hand the network engineer may lack the underlying principles of power and energy systems. A new type of multi-discipline power systems engineer has to be trained. The Smart Grid will soon be a reality. Generation, transmission, distribution consumption and commerce will be information driven. Finally, when automation is combined with mechatronics and robotics, our lives are poised to be drastically changed.

14. References

[1] Mikel P. Groover, Britannica Online Encyclopedia
[2] Benjamin F. Shearer,. "Home front heroes: a biographical dictionary of Americans during wartime", Volume 1, Greenwood Publishing Group 2007
[3] Edward Chikuni, "Concise Higher Electrical Engineering", Juta Academic Publishers, 2008
[4] James Northcote-Green, Robert Wilson, "Control and Automation of Electrical Power Distribution Systems", Taylor& Francis, 2006
[5] AREVA, ALSTOM, Network Protection and Application Guide, 2011 Edition

[6] Su Sheng; Duan Xianzhong; W.L. Chan, *"Probability Distribution of Fault in Distribution System* , Power Systems, IEEE Transactions on, Aug. 2008"

[7] D. Bassett, K. Clinard, J. Grainger, S. Purucker, and D.Ward, *"Tutorial course: distribution automation,"* IEEE Tutorial Publ. 88EH0280-8-PWR, 1988

[8] IEC 61850 Guide Serveron® TM8TM and TM3TM On-line Transformer Monitors 810-1885-00 Rev A August 2011

[9] M.C. Janssen,, A. Apostolov *"IEC 61850 Impact on Substation Design"*

[10] Drew Baigent, Mark Adamiak, Ralph Mackiewicz, GE and SISCO *"Communication Networks and Systems In Substations, An Overview for Users",.*

[11] Karlheinz Schwarz, SCC, *"Monitoring and Control of Power Systems and Communication Infrastructures based on IEC 61850 and IEC 61400".*

[12] Dipl.-Ing. H. Dawidczak, Dr.-Ing. H. Englert Siemens AG, Energy Automation Nuremberg, Germany *"IEC 61850 interoperability and use of flexible object modeling and naming"*

[13] R. Moore, IEEE Member, R. Midence, IEEE, M. Goraj, *"Practical Experience with IEEE 1588 High Precision Time Synchronization in Electrical Substation based on IEC 61850 Process Bus"*

[14] J. Holbach, J. Rodriguez, C. Wester, D. Baigent, L. Frisk, S. Kunsman, L. Hossenlop, *"First IEC 61850 Multivendor Project in the USA, Protection, Automation and Control World, August 2007"*

[15] Edward Chikuni, Maxwell Dondo, *"Investigating the Security of Electrical Power Systems"* SCADA, IEEE Africon 2007, Windhoek

[16] Martin Naedele , Dacfey Dzung , Michael Stanimirov, *"Network Security for Substation Automation Systems"*, Proceedings of the 20th International Conference on Computer Safety, Reliability and Security, p.25-34,September 26-28, 2001

[17] E Chikuni, F Engelbrecht, and M Dondo, "The emergence of substation automation in Southern Africa, opportunities, challenges and threats", IEEE Africon Conference, Windhoek, September, 2007

8

Learning Automation to Teach Mathematics

Josep Ferrer[1], Marta Peña[1] and Carmen Ortiz-Caraballo[2]
[1]*Universitat Politecnica de Catalunya*
[2]*Universidad de Extremadura*
Spain

1. Introduction

One of the recurrent controversies at our Engineering Faculty concerns the orientation of first year basic courses, particularly the subject area of mathematics, considering its role as an essential tool in technological disciplines. In order to provide the basic courses with technological applications, a mathematical engineering seminar was held at the Engineering Faculty of Barcelona. Sessions were each devoted to one technological discipline and aimed at identifying the most frequently used mathematical tools with the collaboration of guest speakers from mathematics and technology departments.

In parallel, the European Space of Higher Education process is presented as an excellent opportunity to substitute the traditional teaching-learning model with another where students play a more active role. In this case, we can use the Problem-Based Learning (PBL) method. This environment is a really useful tool to increase student involvement as well as multidisciplinarity. With PBL, before students increase their knowledge of the topic, they are given a real situation-based problem which will drive the learning process. Students will discover what they need to learn in order to solve the problem, either individually or in groups, using tools provided by the teacher or 'facilitator', or found by themselves.

Therefore, a collection of exercises and problems has been designed to be used in the PBL session of the first course. These exercises include the applications identified in the seminar sessions and would be considered as the real situation-based problems to introduce the different mathematic topics. Two conditions are imposed: availability for first year students and emphasis on the use of mathematical tools in technical subjects in later academic years. As additional material, guidelines for each technological area addressed to faculties without engineering backgrounds are defined.

Some material on Electrical Engineering has been already published ((Ferrer et al., 2010)). Here we focus on control and automation. The guidelines and some exercises will be presented in detail later on. As general references on linear algebra see for example (Puerta, 1976) and on system theory (Kalman et al., 1974) and (Chen, 1984). For other applications, see (Lay, 2007).

Here we describe some of the items regarding control and automation that are presented in the guideline (Section 2).

(1) The input-output description: black box, input-output signals, impulse response, linearity, causality, relaxedness, time invariance, transfer-function matrix, time domain, frequency domain, Laplace transform, Fourier transform, gain, phase, poles, zeros, Bode diagram, filters, resonances.

(2) The state-variable description: the concept of state, state equation, output equation, transfer-function matrix, linear changes, feedbacks, realizations, stability, reachable states, controllability, control canonical form, pole assignment, observability, Kalman decomposition.

The exercises (Section 3) related to this area cover the following topics:

- Matrices. Determinant. Rank: Composition of Systems (ex. 1), Controllability Matrix, Controllable Systems and Controllability Indices (ex. 2), Realizations (ex. 3).

- Vector Spaces. Bases. Coordinates: States in Discrete Systems (ex. 4), Control Functions in Discrete Systems (ex. 4).

- Vector Subspaces: Reachable States for one or several Controls and Sum and Intersection of these Subspaces (ex. 5).

- Linear Maps: Changes of Bases in the System Equations and Invariance of the Transfer-Function Matrix (ex. 6), Controllability Subspace and Unobservable Subspace (ex. 7), Kalman Decomposition (ex. 8).

- Diagonalization. Eigenvectors, Eigenvalues: Invariant Subspaces and Restriction to an Invariant Subspace (ex. 7), Controllable Subsystem (ex. 10), Poles and Pole assignment (ex. 9).

- Non-Diagonalizable Matrices: Control Canonical Form (ex. 11)

2. Guideline for teachers

(1) External system description

Systems are considered as "black boxes" in which each input $u(t)$ (input, control, cause,...) causes an output $y(t)$ (output, effect,...), both multidimensional vectors, in general. We consider only known inputs, ignoring other ones like, for example, disturbance, noise...

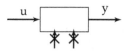

The most usual inputs will be piecewise continuous functions, built from the elementary functions or standard "signals" (impulse or delta, step, ramp, sawtooth, periodics...). Simple systems are adders, gain blocks, integrators, pure delays, filters...

In general, the aim is to analyse their behaviour without looking inside. Indeed, if we consider the "impulse response" $g(t; \tau)$ (that is, the output when the input is an "impulse" in τ, δ_τ) and the system is linear, it results

$$y(t) = \int_{-\infty}^{+\infty} g(t; \tau)u(\tau)d\tau.$$

The upper integration limit will be t if we assume that the system is "causal" (that is, if the current response does not depend on the future inputs, as in all physical systems) and the lower will be t_0 (or simply 0) if it is "relaxed" at t_0 (that is, $y(t) = 0$ for $t \geq t_0$, when $u(t) = 0$ for $t \geq t_0$). Finally, if it is "time-invariant", instead of $g(t; \tau)$ we can write $g(t - \tau)$, resulting in the following expression of $y(t)$ as a convolution product of $g(t)$ and $u(t)$

$$y(t) = \int_{-\infty}^{+\infty} g(t - \tau)u(\tau)d\tau = g(t) * u(t),$$

which is the general system representation in the "time domain".

Applying Laplace transform we get the general representation in the "s-domain"

$$\hat{y}(s) = G(s)\hat{u}(s),$$

where $G(s) = \hat{g}(s)$ is called "transfer function matrix". Indeed, it is the focus of study in this external representation.

If we do the change of variable $s = j\omega$ (imaginary axis of the "complex plane") we obtain the representation in the "frequency domain", (more usual in engineering)

$$y(\omega) = G(\omega)u(\omega),$$

where $G(\omega)$ is called "isochronous transfer function matrix". It can be obtained directly as a "Fourier" transform when appropriate hypotheses hold. This allows the use of tools such as Fourier transform, Parseval theorem... if we have some basic knowledge of functional analysis and complex variable: function spaces, norms, Hilbert spaces, integral transforms...

Its denomination shows that $G(\omega)$ indicates the system behaviour for each frequency. So, if $u(t) = A\sin(\omega t)$, then $y(t) = B\sin(\omega t + \varphi)$, being the "gain" B/A the module of $G(\omega)$, and the "phase" φ is its argument. A widely used tool in engineering is the "Bode diagrams" which represent these magnitudes on the ordinate (usually, the gain is in logarithmic scale, in decibels: $dB = 20log|G|$) as a function of the frequency as abscissa (also in log scale).

In generic conditions (see next section) the coefficients of G are "proper rational fractions", so that the system's behaviour is largely determined by its "degrees", "zeros" and "poles". So:

- The relative degree (denominator degree minus numerator degree) gives the "order of differentiability" at the origin of the response to the input step signal.

- As already mentioned, this difference must be strictly positive (if zero, the Parseval theorem would give infinite energy in the output signal; if negative, it would contradict the causality), so that the gain tends to 0 for high frequencies.

- Roughly speaking, the zeros indicate filtered frequencies (of gain 0, or in practice far below), so that in the Bode diagram place the "inverted comb pas" as "filters". On the contrary, the poles indicate dangerous frequencies because of resonance (infinite gain).

(2) Internal system description

In addition, state variables $x(t)$ (not univocally defined) are considered. They characterize the state in the sense that they accumulate all the information from the past, that is, future outputs are determined by the current state and the future inputs. Typically, the derivative $\dot{x}(t)$

functionally depends on $x(t)$, $u(t)$ and t (state equation), and $y(t)$ as well (output equation), although in this case we may obviate the dependence on $u(t)$. In the linear case:

$$\dot{x}(t) = A(t)x(t) + B(t)u(t); \quad y(t) = C(t)x(t).$$

From elementary theory of ordinary differential equations, it holds that for every continuous (or piecewise continuous) control $u(t)$, there exists a unique "solution" $x(t)$ for every "initial condition" $x_0 = x(0)$.

In particular, it can be applied a "feedback control" by means of a matrix F

$$u(t) = Fx(t).$$

One of the first historical (non linear) examples is the Watt regulator which controlled the velocity of a steam engine acting on the admission valve in the function of the centrifugal force created in the regulator balls by this velocity. Nowadays this "automatic regulation" can be found in simple situations such as thermostats.

When this kind of feedback is applied, we obtain an autonomous dynamical system

$$\dot{x}(t) = Ax(t) + BFx(t) = (A + BF)x(t)$$

called a "closed loop" system. It is natural to consider if we can adequately choose F such that this system has suitable dynamic properties. For example, for being "stable", that is, that the real parts of the eigenvalues of $A + BF$ are negative. Or more in general, that these eigenvalues have some prefixed values. As we will see later, one of the main results is that this feedback "pole assignment" is possible if the initial system is "controllable".

If it is time-invariant (that is, A, B, and C are constant) and we assume $x(0) = 0$, the Laplace transform gives

$$\hat{y}(t) = G(s)\hat{u}(s); \quad G(s) = C(sI - A)^{-1}B$$

recovering the previous transfer matrix. Reciprocally, the "realization" theory constructs triples (A, B, C) giving a prefixed G, formed by proper rational fractions: it can be seen that it is always possible; the uniqueness conditions will be seen later.

It is a simple exercise to check that when introducing a "linear change" S in the state variables, the system matrices become $S^{-1}AS$, $S^{-1}B$ and CS, respectively, and $G(s)$ do not change.

In this description it is clear that the coefficients of $G(s)$ are proper rational fractions and in particular its poles are the eigenvalues of A.

One of the main results is that the set of "reachable states" (the possible $x(t)$ starting from the origin, when varying the controls $u(t)$) is the image subspace $\operatorname{Im} K(A, B)$ of the so-called "controllability matrix"

$$\operatorname{Im} K(A,B); \quad K(A,B) = \begin{pmatrix} B & AB & \ldots & A^{n-1}B \end{pmatrix},$$

that is, the subspace spanned by the columns of B and successive images for A, called "controllability subspace". It is an interesting exercise to justify that we can truncate in $n - 1$, since the columns of higher powers are linearly dependent with the previous ones. In fact we can consider each control individually (each column of B separately), in such a way that the "sum and intersection of subspaces" give the reachable states when using different controls at the same time or for each one of them.

The system is "controllable" if all the states are reachable, that is, if and only if $K(A, B)$ has full "rank". Hence, it is a generic condition (that is, the subset of controllable pairs (A, B) is open and dense).

In the single input case (A, b), it can be seen that if the system is controllable it can be transformed, by means of a suitable basis change, in the so-called "control canonical form"

$$\bar{A} = \begin{pmatrix} 0 & 1 & 0 & \dots & 0 & 0 \\ 0 & 0 & 1 & \dots & 0 & 0 \\ & & \dots & \dots & & \\ 0 & 0 & 0 & \dots & 0 & 1 \\ * & * & * & \dots & * & * \end{pmatrix}, \quad \bar{b} = \begin{pmatrix} 0 \\ 0 \\ \vdots \\ 0 \\ 1 \end{pmatrix}.$$

Observe that $\bar{A}$ is a "companion" (or Sylvester) matrix. For these kinds of matrices, it can be easily seen that the coefficients of the characteristic polynomial are the ones of the last row (with opposite sign and in reverse order), and they are "non derogatory matrices", that is, for each eigenvalue there is a unique linearly independent eigenvector, and therefore only one Jordan block. Hence, if some eigenvalue is multiple, the matrix is non diagonalizable.

In the multi input case, another reduced form is used (Brunovsky, or Kronecker form), which is determined by the so-called "controllability indices" that can be computed as a conjugated partition from the one of the ranks of: $B, (B, AB),...,K(A, B)$.

From these reduced forms it is easy to prove that the pole assignment is feasible, as well as to compute the suitable feedbacks.

If the system is not controllable, it is easy to see that the subspace Im $K(A, B)$ is "A-invariant", and that in any "adapted basis" (of the state space) the matrices of the system are of the form:

$$\begin{pmatrix} A_c & * \\ 0 & * \end{pmatrix}, \quad \begin{pmatrix} B_c \\ 0 \end{pmatrix},$$

being (A_c, B_c) controllable. It is also easy to deduce that if $x(0)$ belongs to Im $K(A, B)$, $x(t)$ also belongs, for all time t and all control $u(t)$, and that its trajectory is determined for the pair (A_c, B_c), which enables considering the "restriction" of the system (A, B) to Im $K(A, B)$, called the "controllable subsystem" of the initial one.

In the same direction, the "Kalman decomposition" is obtained by considering "(Grassman) adapted bases to the pair of subspaces" Im $K(A, B)$ and Ker $L(C, A)$, being

$$L(C, A) = \begin{pmatrix} C \\ CA \\ \dots \\ CA^{n-1} \end{pmatrix}$$

that is, the "transposed matrix" of $K(A^t, C^t)$. In fact there are interesting properties of "duality" between the systems (A, B, C) and (A^t, C^t, B^t).

We remark the equivalence between the controllability of (A, B, C) and the "observability" of (A^t, C^t, B^t), in the sense that the initial conditions are computable if the outputs for determined inputs are known.

On the other hand it is interesting to note that the transfer matrix of the initial system is the same as that of the controllable subsystem, as well as that of the "complete subsystem" (that is, controllable and observable) obtained by means of the Kalman decomposition.

This means that, given any transfer matrix (formed by proper rational fractions), not only it is possible to find realizations, but also "controllable realizations" and even complete ones. In fact it can be seen that all the complete realizations are equivalent. In particular, they have the same number of state variables. Moreover, a realization is complete if and only if it is a "minimal realization", in the sense that there are not realizations with a smaller number of state variables. This minimal number of state variables of the realizations is called "McMillan degree", which coincides with the dimension of the complete realizations.

3. Exercises for students

We will present here the guideline of proposed exercises for the students and their solutions:

3.1 Proposed exercises

1. Composition of systems

A control system Σ

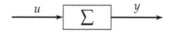

is defined by the equations

$$\dot{x}(t) = Ax(t) + Bu(t); \qquad y(t) = Cx(t),$$

or simply by the "triple of matrices" (A, B, C). This triple determines the "transfer matrix"

$$G(s) = C(sI - A)^{-1}B$$

which relates the Laplace transforms of u, y:

$$\hat{y}(s) = G(s)\hat{u}(s).$$

Given two systems:

$$\Sigma_1 : \quad \dot{x}_1 = A_1x_1 + B_1u_1; \quad y_1 = C_1x_1$$
$$\Sigma_2 : \quad \dot{x}_2 = A_2x_2 + B_2u_2; \quad y_2 = C_2x_2$$

they can be composed of different ways to obtain a new system. For example in:

(i) Series

(ii) Parallel

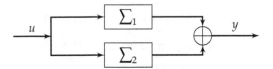

(iii) Feedback

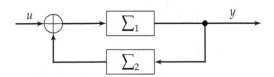

In all these cases the new state variables are:

$$x = \begin{pmatrix} x_1 \\ x_2 \end{pmatrix}$$

(1) Deduce the following relations between the different input and output variables:

(1.i) $u = u_1; y_1 = u_2; y_2 = y$

(1.ii) $u = u_1 = u_2; y = y_1 + y_2$

(1.iii) $u + y_2 = u_1; y_1 = y = u_2$

(2) Deduce the expression of the triple of matrices of each composed system, in terms of $A_1, B_1, C_1, A_2, B_2, C_2$.

(3) Deduce the expression of the transfer matrix of each composed system, in terms of $G_1(s)$ and $G_2(s)$.

2. Controllable systems; controllability indices

A control system

$$\dot{x}(t) = Ax(t) + Bu(t), \quad A \in M_n, \quad B \in M_{n,m}$$

is "controllable" (that is, any change in the values of x is possible by means of a suitable control $u(t)$), if and only if the rank of the so-called "controllability matrix" is full

$$K(A, B) = (B, AB, A^2B, \ldots, A^{n-1}B).$$

(a) Discuss for which values of $\alpha, \beta \in \mathbb{R}$ the system defined by

$$A = \begin{pmatrix} 0 & 1 & 0 \\ -2 & 1 & 0 \\ \alpha & 0 & 1 \end{pmatrix}, \quad B = \begin{pmatrix} 1 & 0 \\ 0 & \beta \\ 0 & 1 \end{pmatrix}$$

is controllable.

(b) Discuss for which values of $\alpha, \beta \in \mathbb{R}$ the system is controllable with only the second control, that is, when instead of the initial matrix B we consider only the column matrix

$$b_2 = \begin{pmatrix} 0 \\ \beta \\ 1 \end{pmatrix}.$$

(c) In general, the "controllability indices" are determined by the rank of the matrices

$$B, \quad (B, AB), \quad (B, AB, A^2B), \dots, \quad K(A, B).$$

Compute these ranks, in terms of parameters $\alpha, \beta \in \mathbb{R}$, for the system defined by

$$A = \begin{pmatrix} 0 & 1 & & & & \\ & 0 & 1 & & & \\ & & 0 & 1 & & \\ & & & 0 & 1 & 0 \\ & & & & 0 & \\ \hline & & 0 & & 0 & 1 \\ & & & & & 0 \end{pmatrix}, \quad B = \begin{pmatrix} 0 & 0 \\ 0 & 0 \\ 0 & \gamma \\ 0 & \delta \\ 1 & 0 \\ \hline 0 & 0 \\ 0 & 1 \end{pmatrix}.$$

3. Realizations

Given a matrix $G(s)$ of proper rational fractions in the variable s, it is called a "realization" any linear control system

$$\dot{x}(t) = Ax(t) + Bu(t); \quad y(t) = Cx(t)$$

which has $G(s)$ as "transfer matrix", that is,

$$G(s) = C(sI - A)^{-1}B.$$

In particular, the so-called "standard controllable realization" is obtained in the following way, assuming that $G(s)$ has p rows and m columns.

(i) We determine the least common multiple polynomial of the denominators

$$P(s) = p_0 + p_1 s + \dots + p_r s^r.$$

(ii) Then, $G(s)$ can be written

$$G(s) = \frac{1}{P(s)} \begin{pmatrix} G_{11}(s) & \cdots & G_{1m}(s) \\ \cdots & \cdots & \cdots \\ G_{p1}(s) & \cdots & G_{pm}(s) \end{pmatrix},$$

where $G_{ij}(s)$ are polynomials with degree strictly lower than $r(= \deg P(s))$.

(iii) Grouping the terms of the same degree we can write $G(s)$ in the form

$$G(s) = \frac{1}{P(s)}(R_0 + R_1 s + \dots + R_{r-1} s^{r-1}),$$

where $R_0, \dots, R_{r-1} \in M_{p,m}(\mathbb{R})$.

(iv) Then, the standard controllable realization is given by the triple of matrices (where 0_m is the null matrix of $M_m(\mathbb{R})$):

$$A = \begin{pmatrix} 0_m & I_m & 0_m & \cdots & 0_m \\ 0_m & 0_m & I_m & \cdots & 0_m \\ & \cdots & \cdots\cdots & & \\ 0_m & 0_m & 0_m & \cdots & I_m \\ -p_0 I_m & -p_1 I_m & \cdots\cdots & & -p_{r-1}I_m \end{pmatrix} \in M_{mr}(\mathbb{R}), \quad B = \begin{pmatrix} 0_m \\ \vdots \\ 0_m \\ I_m \end{pmatrix} \in M_{mr,m}(\mathbb{R}),$$

$$C = \begin{pmatrix} R_0 & \cdots & R_{r-1} \end{pmatrix} \in M_{p,mr}(\mathbb{R}).$$

We consider, for example,

$$G(s) = \begin{pmatrix} 1/s \\ 1/(s-1) \end{pmatrix}.$$

(1) Following the above paragraphs, compute the triple of matrices (A, B, C) which give the standard controllable realization.

(2) Check that it is controllable, that is, that the matrix

$$K(A, B) = (B, AB, \ldots, A^{mr-1}B)$$

has full rank.

(3) Check that it is a realization of $G(s)$, that is,

$$\begin{pmatrix} 1/s \\ 1/(s-1) \end{pmatrix} = C(sI - A)^{-1}B.$$

4. Reachable states; control functions

Given a linear control system

$$x(k+1) = Ax(k) + Bu(k); \quad A \in M_n(\mathbb{R}), \quad B \in M_{n,m}(\mathbb{R}),$$

the h-step "reachable states", from $x(0)$, are:

$$x(h) = A^h x(0) + A^{h-1}Bu(0) + \ldots + ABu(h-2) + Bu(h-1)$$

where all possible "control functions" $u(0), u(1), \ldots$ are considered. More explicitly, if we write $B = (b_1 \ldots b_m)$ and $u(k) = (u_1(k), \ldots, u_m(k))$, we have that:

$$x(h) = A^h x(0)$$
$$+ (A^{h-1}b_1 u_1(0) + \ldots + A^{h-1}b_m u_m(0)) + \ldots$$
$$+ (Ab_1 u_1(h-2) + \ldots + Ab_m u_m(h-2)) + (b_1 u_1(h-1) + \ldots + b_m u_m(h-1))$$

Let

$$A = \begin{pmatrix} 2 & -1 & 1 \\ -2 & 1 & -1 \\ 2 & 1 & 3 \end{pmatrix}, \quad B = \begin{pmatrix} -1 & 1 \\ 1 & 0 \end{pmatrix}.$$

(1) Assume that only the first control acts, that is: $u_2(0) = u_2(1) = \ldots = 0$. Show that in this case, the state $x = (-1, 2, 1)$ is not 3-step reachable from the origin.

(2) Assume now that only the second control acts.

    (2.1) Prove that $x = (-1, 2, 1)$ is 2-step reachable from the origin.

    (2.2) Compute the corresponding control function.

(3) Assume that both controls act.

    (3.1) Determine the control functions set to reach $x = (-1, 2, 1)$ from the origin, at second step.

    (3.2) In particular, check if it is possible to choose positive controls.

    (3.3) Idem for $u_1(0) = u_2(0)$.

5. Subspaces of reachable states

Given a linear control system

$$x(k+1) = Ax(k) + Bu(k); \quad A \in M_n(\mathbb{R}), \quad B \in M_{n,m}(\mathbb{R}),$$

the h-step "reachable states", from $x(0)$, are

$$x(h) = A^h x(0) + A^{h-1} Bu(0) + \ldots + ABu(h-2) + Bu(h-1),$$

where all possible control functions $u(0), u(1), \ldots$ are considered.

(1) We write $K(h)$ the set of these states when $x(0) = 0$. Show that:

    (1.1) $K(h) = [B, AB, \ldots, A^{h-1}B] \subset \mathbb{R}^n$

    (1.2) $K(1) \subset K(2) \subset \ldots \subset K(j) \subset \ldots$

    (1.3) $K(h) = K(h+1) \Rightarrow K(h+1) = K(h+2) = \ldots$

    (1.4) $K(n) = K(n+1) = \ldots$

This maximal subspace of the chain is called "subspace of reachable states":

$$K = [B, AB, \ldots, A^{n-1}B] \subset \mathbb{R}^n$$

(2) Analogous results hold when only the control $u_i(k)$ acts. In particular, the subspace of reachable states, from the origin, with only this control is:

$$K_i = [b_i, Ab_i, \ldots, A^{n-1}b_i] \subset \mathbb{R}^n; \quad 1 \le i \le m$$

where $B = (b_1, \ldots, b_m)$.

    (2.1) Reason that the reachable states by acting the controls $u_i(k)$ and $u_j(k)$ are $K_i + K_j$.

    (2.2) Reason that $K_i \cap K_j$ is the subspace of reachable states by acting any of the controls $u_i(k)$ or $u_j(k)$.

(3) Let us consider the linear control system defined by the matrices

$$A = \begin{pmatrix} 2 & -1 & 1 \\ -2 & 1 & -1 \\ 2 & 1 & 3 \end{pmatrix}, \quad B = \begin{pmatrix} 0 & 0 \\ -1 & 1 \\ 1 & 0 \end{pmatrix}.$$

(3.1) Determine the subspaces K, K_1 and K_2, and construct a basis of each one.

(3.2) Idem for $K_1 + K_2$, $K_1 \cap K_2$.

6. Change of state variables in control systems

In the linear control system

$$\dot{x}(t) = Ax(t) + Bu(t); \quad y(t) = Cx(t)$$

$$A \in M_n, \quad B \in M_{n,m}, \quad C \in M_{p,n},$$

we consider a linear change in the state variables given by:

$$\bar{x} = S^{-1}x.$$

(1) Prove that in the new variables the equations of the system are:

$$\dot{\bar{x}}(t) = \bar{A}\bar{x}(t) + \bar{B}\bar{u}(t); \quad y(t) = \bar{C}\bar{x}(t)$$

$$\bar{A} = S^{-1}AS, \quad \bar{B} = S^{-1}B, \quad \bar{C} = CS.$$

(2) The "controllability indices" of the system are computed from the ranks:

$$\text{rank}(B, AB, \ldots, A^hB), \quad h = 1, 2, 3, \ldots$$

Deduce from (1) that they are invariant under linear changes in the state variables.

(3) The "transfer matrix" of the system is:

$$G(s) = C(sI - A)^{-1}B.$$

Deduce from (1) that it is invariant under linear changes in the state variables.

7. Controllability subspaces and unobservability subspaces

Given a linear control system

$$\dot{x}(t) = Ax(t) + Bu(t); \quad y(t) = Cx(t)$$

$$A \in M_n(\mathbb{R}), \quad B \in M_{n,m}(\mathbb{R}), \quad C \in M_{p,n}(\mathbb{R}),$$

the following subspaces are called "controllability subspace" and "unobservability subspace", respectively:

$$K = \text{Im}\begin{pmatrix} B & AB & \ldots & A^{n-1}B \end{pmatrix}$$

$$L = \text{Ker}\begin{pmatrix} C \\ CA \\ \ldots \\ CA^{n-1} \end{pmatrix}.$$

(1) Show that they are A-invariant subspaces.

(2) Let us consider

$$A = \begin{pmatrix} 1 & 1 & 0 & 0 \\ 0 & 1 & 0 & 0 \\ 0 & 0 & -1 & 1 \\ 0 & 0 & 0 & -1 \end{pmatrix}, \quad B = \begin{pmatrix} 1 \\ 0 \\ 0 \\ 1 \end{pmatrix}, \quad C = (0\ 1\ 0\ 1).$$

(2.1) Compute the dimensions of K and L.

(2.2) Construct a basis of each one.

(2.3) Obtain the matrices, in these bases, of the restrictions $A|_K$ and $A|_L$.

(2.4) Idem for the subspace $K \cap L$.

8. Kalman decomposition

Given a linear control system

$$\dot{x}(t) = Ax(t) + Bu(t); \quad y(t) = Cx(t)$$

$$A \in M_n(\mathbb{R}), \quad B \in M_{n,m}(\mathbb{R}), \quad C \in M_{p,n}(\mathbb{R}),$$

we consider the subspaces

$$K = \operatorname{Im} K(A, B), \quad L = \operatorname{Ker} L(C, A),$$

where

$$K(A, B) = (\, B \ AB \ \dots \ A^{n-1}B \,),$$

$$L(C, A) = \begin{pmatrix} C \\ CA \\ \dots \\ CA^{n-1} \end{pmatrix}.$$

A (Grassman) adapted basis to both subspaces is called a "Kalman basis". More specifically, the basis change matrix is of the form

$$S = (\, S_1 \ S_2 \ S_3 \ S_4 \,)$$
$$S_2 : \text{ basis of } K \cap L$$
$$(S_1 \quad S_2) : \text{ basis of } K$$
$$(S_2 \quad S_4) : \text{ basis of } L$$

where some of the submatrices $S_1, \dots, S_4$ can be empty.

(1) Prove that with a basis change of this form, the matrices of the system become of the form

$$\bar{A} = S^{-1}AS = \begin{pmatrix} A_{11} & 0 & A_{13} & 0 \\ A_{21} & A_{22} & A_{23} & A_{24} \\ 0 & 0 & A_{33} & 0 \\ 0 & 0 & A_{43} & A_{44} \end{pmatrix}$$

$$\bar{B} = S^{-1}B = \begin{pmatrix} B_1 \\ B_2 \\ 0 \\ 0 \end{pmatrix}$$

$$\bar{C} = CS = (C_1 \ 0 \ C_3 \ 0).$$

It is called "Kalman decomposition" of the given system.

(2) Consider the system given by

$$A = \begin{pmatrix} 1 & 1 & 0 & 0 \\ 0 & 1 & 0 & 0 \\ 0 & 0 & -1 & 1 \\ 0 & 0 & 0 & -1 \end{pmatrix}, \quad B = \begin{pmatrix} 1 \\ 0 \\ 0 \\ 1 \end{pmatrix}, \quad C = (0 \ 1 \ 0 \ 1).$$

(2.1) Determine a Kalman basis.

(2.2) Determine a Kalman decomposition.

(2.3) Check that

(A_{11}, B_1, C_1) is controllable and observable,

$\left(\begin{pmatrix} A_{11} & 0 \\ A_{21} & A_{22} \end{pmatrix}, \begin{pmatrix} B_1 \\ B_2 \end{pmatrix}, \begin{pmatrix} C_1 \\ 0 \end{pmatrix} \right)$ is controllable,

$\left(\begin{pmatrix} A_{11} & A_{13} \\ 0 & A_{33} \end{pmatrix}, \begin{pmatrix} B_1 \\ 0 \end{pmatrix}, \begin{pmatrix} C_1 \\ C_3 \end{pmatrix} \right)$ is observable,

that is, that the following matrices have full rank:

$$K(A_{11}, B_1) = (B_1 \ A_{11}B_1 \ \dots \ (A_{11})^{n-1}B_1),$$

$$(L(C_1, A_{11}))^t = (C_1^t \ A_{11}^t C_1^t \ \dots \ (A_{11}^t)^{n-1}C_1^t),$$

$$K\left(\begin{pmatrix} A_{11} & 0 \\ A_{21} & A_{22} \end{pmatrix}, \begin{pmatrix} B_1 \\ B_2 \end{pmatrix} \right) = \left(\begin{pmatrix} B_1 \\ B_2 \end{pmatrix} \dots \begin{pmatrix} A_{11} & 0 \\ A_{21} & A_{22} \end{pmatrix}^{n-1} \begin{pmatrix} B_1 \\ B_2 \end{pmatrix} \right),$$

$$\left(L\left((C_1 \ C_3), \begin{pmatrix} A_{11} & A_{13} \\ 0 & A_{33} \end{pmatrix} \right) \right)^t = \left(\begin{pmatrix} C_1^t \\ C_3^t \end{pmatrix} \dots \begin{pmatrix} A_{11}^t & 0 \\ A_{13}^t & A_{33}^t \end{pmatrix}^{n-1} \begin{pmatrix} C_1^t \\ C_3^t \end{pmatrix} \right).$$

9. Pole assignment by state feedback

Given a linear control system

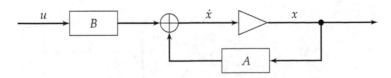

$$\dot{x}(t) = Ax(t) + Bu(t); \quad A \in M_n, \quad B \in M_{n,m},$$

the eigenvalues of the matrix A are called the "poles" of the system and play an important role in its dynamic behaviour. For example, they are the resonance frequencies; the system is "BIBO stable" if and only if the real part of its poles is negative.

If we apply an automatic control by means of a "feedback" F (for example, a thermostat, a Watt regulator,...)

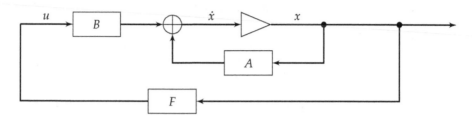

we obtain an autonomous dynamic system (called a "closed loop" system):

$$\dot{x}(t) = Ax(t) + BFx(t) = (A + BF)x(t)$$

with matrix $A + BF$ instead of the initial matrix A. We ask if it is possible to choose adequately F such that the new poles (that is, the eigenvalues of $A + BF$) have some prefixed desired values. For example, with negative real part, so that the automatic control is stable.

A main result of control theory ensures that this pole assignment by feedback is possible if the initial system is controllable, that is, if $\text{rank}(B, AB, A^2B, \ldots, A^{n-1}B) = n$.

We will prove and apply that in the one-parameter case ($m = 1$). We will do it by transforming the initial system in the so-called "control canonical form" by means of a suitable basis change. In this form, the feedback computation is trivial, and finally it will only be necessary to undo the transformation.

(1) Consider a one-parameter control system

$$\dot{x}(t) = Ax(t) + bu(t); \quad A \in M_n, \quad b \in M_{n,1},$$

and let $\mu_1, \ldots, \mu_n$ be the desired eigenvalues for the feedback system (each one repeated so many times as its algebraic multiplicity). The main hypothesis is that the system is controllable, that is, $\text{rank}(b, Ab, A^2b, \ldots, A^{n-1}b) = n$.

Then it is possible to find a linear change S in the state variables such that the new matrices of the system have the so-called "control canonical form":

$$A_c = S^{-1}AS = \begin{pmatrix} 0 & 1 & 0 & \ldots & 0 & 0 \\ 0 & 0 & 1 & \ldots & 0 & 0 \\ & \ldots & & \ldots & & \\ 0 & 0 & 1 & \ldots & 0 & 1 \\ a_n & a_{n-1} & a_{n-2} & \ldots & a_2 & a_1 \end{pmatrix}, \quad b_c = S^{-1}b = \begin{pmatrix} 0 \\ \vdots \\ 0 \\ 1 \end{pmatrix}$$

for certain coefficients $a_1, \ldots, a_n$.

(1.1) Check that these coefficients are the same (with opposite sign) than the ones of the characteristic polynomial:

$$Q(t) = (-1)^n(t^n - a_1 t^{n-1} - \ldots - a_{n-1}t - a_n).$$

(1.2) Deduce that it is straightforward to find F_c such that:

$$\text{eigenvalues}(A_c + bcF_c) = \{\mu_1, \ldots, \mu_n\}.$$

(1.3) Prove that $F = F_c S^{-1}$ is the sought feedback, that is, that:

$$\text{eigenvalues}(A + bF) = \{\mu_1, \ldots, \mu_n\}.$$

(2) Consider the particular case:

$$A = \begin{pmatrix} 0 & 0 & 1 \\ 0 & 1 & -1 \\ 1 & 0 & -1 \end{pmatrix}, \quad b = \begin{pmatrix} 1 \\ 1 \\ 1 \end{pmatrix}.$$

(2.1) Check that the basis change

$$S = \begin{pmatrix} -2 & 1 & 1 \\ -2 & 0 & 1 \\ -1 & 0 & 1 \end{pmatrix}$$

transform the initial matrices into the control canonical form.

(2.2) Compute a feedback F such that:

$$\text{eigenvalues}(A + bF) = \{\mu_1, \mu_2, \mu_3\}.$$

10. Controllable subsystem

Given a linear control system

$$\dot{x}(t) = Ax(t) + Bu(t); \quad y(t) = Cx(t) \tag{1}$$
$$A \in M_n, \quad B \in M_{n,m}, \quad C \in M_{p,n},$$

the following matrix is called its "controllability matrix"

$$K(A, B) = \begin{pmatrix} B & AB & A^2B & \ldots & A^{n-1}B \end{pmatrix}.$$

It can be seen that the subspace spanned by its columns $K = \text{Im}\, K(A, B)$ is the set of reachable states from the origin, and it is called "controllability subspace". We denote $d = \dim K$.

(1) Prove that K is an A-invariant subspace.

We consider a basis change S in the state space adapted to K, and we denote by x_c the first d new coordinates and by x_{uc} the remaining ones:

$$\tilde{x} = S^{-1}x = \begin{pmatrix} x_c \\ x_{uc} \end{pmatrix}, \quad x_c \in M_{d,1}.$$

(2) Check that:

$$x \in K \Leftrightarrow x_{uc} = 0$$

that is:

$$x \in K \Leftrightarrow \bar{x} = S^{-1}x = \begin{pmatrix} x_c \\ 0 \end{pmatrix}.$$

(3) Prove that the equations of the system in the new variables are:

$$\begin{pmatrix} \dot{x}_c(t) \\ \dot{x}_{uc}(t) \end{pmatrix} = \bar{A} \begin{pmatrix} x_c(t) \\ x_{uc}(t) \end{pmatrix} + \bar{B}u(t)$$

$$\bar{A} = S^{-1}AS = \begin{pmatrix} A_c & * \\ 0 & * \end{pmatrix}, \quad A_c \in M_d$$

$$\bar{B} = S^{-1}B = \begin{pmatrix} B_c \\ 0 \end{pmatrix}, \quad B_c \in M_{d,m}$$

$$\bar{C} = CS = \begin{pmatrix} C_c \\ C_{uc} \end{pmatrix}, \quad C_c \in M_{p,d}.$$

(4) Deduce that if the initial state belongs to K, it also belongs all the trajectory, for any applied control $u(t)$:

$$x(0) \in K \Rightarrow x(t) \in K, \quad \forall t, \quad \forall u(t).$$

Therefore, it makes sense to consider the "restriction" to K of the initial system:

$$\dot{x}_c(t) = A_c x_c(t) + B_c u(t); \quad y_c(t) = C_c x_c(t). \tag{2}$$

The following paragraphs justify the interest of this subsystem and its denomination as "controllable subsystem" of the initial one.

(5) Justify that the trajectories of system (1) in K can be computed by means of equations (2) of the subsystem and relation (2).

(6) Prove that this subsystem is "controllable", that is, that the following matrix has full rank:

$$\begin{pmatrix} B_c & A_c B_c & \dots & A_c^{d-1} B_c \end{pmatrix}.$$

(7) Prove that the controllable subsystem has the same "transfer matrix" (which, we recall, reflects the input/output behaviour) than the initial system, that is, that:

$$C(sI - A)^{-1}B = C_c(sI - A_c)^{-1}B_c.$$

11. Control canonical form

For controllable systems, the so-called "control canonical form" simplifies the computations, for example, for the pole assignment by feedback. We are going to obtain it for the one-parameter case.

(1) Consider the system

$$\dot{x}(t) = Ax(t) + bu(t)$$

$$A = \begin{pmatrix} 0 & 0 & -1 \\ 0 & 1 & -1 \\ 1 & 0 & 1 \end{pmatrix}, \quad b = \begin{pmatrix} 1 \\ 1 \\ 1 \end{pmatrix}.$$

(1.1) Check that it is controllable, that is, that:

$$\text{rank}\,(\,b \ Ab \ A^2b\,) = 3.$$

(1.2) Check that doing the basis change

$$S_1 = (\,A^2b \ Ab \ b\,)$$

we get

$$\bar{A} = S_1^{-1}AS_1 = \begin{pmatrix} 2 & 1 & 0 \\ 0 & 0 & 1 \\ -1 & 0 & 0 \end{pmatrix}, \quad \bar{b} = S^{-1}b = \begin{pmatrix} 0 \\ 0 \\ 1 \end{pmatrix}.$$

(1.3) Check that with the additional change

$$S_2 = \begin{pmatrix} 1 & 0 & 0 \\ -2 & 1 & 0 \\ 0 & -2 & 1 \end{pmatrix}$$

we get the control canonical form

$$A_c = S_2^{-1}\bar{A}S_2 = \begin{pmatrix} 0 & 1 & 0 \\ 0 & 0 & 1 \\ -1 & 2 & 0 \end{pmatrix}, \quad b_c = S_2^{-1}\bar{b} = \begin{pmatrix} 0 \\ 0 \\ 1 \end{pmatrix}.$$

(2) Consider the system

$$\dot{x}(t) = Ax(t) + bu(t); \quad A \in M_n, \quad b \in M_{n,1}$$

which is assumed to be controllable:

$$\text{rank}\,(\,b \ Ab \ \dots \ A^{n-1}b\,) = n$$

and let $Q(t)$ be the characteristic polynomial of A

$$Q(t) = (-1)^n(t^n - a_1t^{n-1} - \dots - a_{n-1}t - a_n).$$

(2.1) Check that doing the basis change

$$S_1 = (\,A^{n-1}b \ \dots \ Ab \ b\,)$$

we get

$$\bar{A} = S_1^{-1}AS_1 = \begin{pmatrix} a_1 & 1 & 0 & \dots & 0 \\ a_2 & 0 & 1 & \dots & 0 \\ & & \dots & & \\ a_{n-1} & 0 & 0 & \dots & 1 \\ a_n & 0 & 0 & \dots & 0 \end{pmatrix}, \quad \bar{b} = S_1^{-1}b = \begin{pmatrix} 0 \\ \vdots \\ 0 \\ 1 \end{pmatrix}.$$

(2.2) Check that with the additional change

$$S_2 = \begin{pmatrix} 1 & 0 & \dots & 0 & 0 \\ -a_1 & 1 & \dots & 0 & 0 \\ -a_2 & -a_1 & \dots & 0 & 0 \\ & & \dots & & \\ -a_{n-1} & -a_{n-2} & \dots & -a_1 & 1 \end{pmatrix}$$

we get the control canonical form

$$A_c = S_2^{-1} \tilde{A} S_2 = \begin{pmatrix} 0 & 1 & 0 & \dots & 0 & 0 \\ 0 & 0 & 1 & \dots & 0 & 0 \\ & & & \dots & & \\ 0 & 0 & 0 & \dots & 0 & 1 \\ a_n & a_{n-1} & a_{n-2} & \dots & a_2 & a_1 \end{pmatrix}, \quad b_c = S_2^{-1} \bar{b} = \begin{pmatrix} 0 \\ \vdots \\ 0 \\ 1 \end{pmatrix}.$$

(3) As an application, we use the control canonical form to check that A is non-derogatory:

(3.1) Prove that if λ_i is an eigenvalue of A, then:

$$\text{rank}(A_C - \lambda_i I) = n - 1.$$

(3.2) Deduce that A is non-derogatory, and so it does not diagonalize if some of its eigenvalues are multiple.

3.2 Solutions

1. Solution

(1) It follows immediately from the observation of the diagrams.

(2.i)

$$\dot{x}_1 = A_1 x_1 + B_1 u_1 = A_1 x_1 + B_1 u$$
$$\dot{x}_2 = A_2 x_2 + B_2 u_2 = A_2 x_2 + B_2 y_1 = A_2 x_2 + B_2 C_1 x_1$$
$$y = y_2 = C_2 x_2$$

Hence:

$$\dot{x} = \begin{pmatrix} \dot{x}_1 \\ \dot{x}_2 \end{pmatrix} = \begin{pmatrix} A_1 & 0 \\ B_2 C_1 & A_2 \end{pmatrix} \begin{pmatrix} x_1 \\ x_2 \end{pmatrix} + \begin{pmatrix} B_1 \\ 0 \end{pmatrix} u$$
$$y = \begin{pmatrix} 0 & C_2 \end{pmatrix} \begin{pmatrix} x_1 \\ x_2 \end{pmatrix}$$

(2.ii) Reasoning in the same way, it results:

$$\begin{pmatrix} A_1 & 0 \\ 0 & A_2 \end{pmatrix}, \quad \begin{pmatrix} B_1 \\ B_2 \end{pmatrix}, \quad \begin{pmatrix} C_1 & C_2 \end{pmatrix}$$

(2.iii) Analogously:

$$\begin{pmatrix} A_1 & B_1 C_2 \\ B_2 C_1 & A_2 \end{pmatrix}, \quad \begin{pmatrix} B_1 \\ 0 \end{pmatrix}, \quad (C_1 \; 0)$$

(3.i) $\hat{y} = \hat{y}_2 = G_2 \hat{u}_2 = G_2 \hat{y}_1 = G_2 G_1 \hat{u}_1 = G_2 G_1 \hat{u}$. Hence, the transfer matrix is:

$$G_2 G_1$$

(3.ii) Analogously:

$$G_1 + G_2$$

(3.iii) $\hat{y} = \hat{y}_1 = G_1 \hat{u}_1 = G_1 (\hat{u} + \hat{y}_2) = G_1 \hat{u} + G_1 G_2 \hat{u}_2 = G_1 \hat{u} + G_1 G_2 \hat{y}$

$(I - G_1 G_2) \hat{y} = G_1 \hat{u}$

$\hat{y} = (I - G_1 G_2)^{-1} G_1 \hat{u}$

2. Solution

(a) The controllability matrix is

$$K(A,B) = \begin{pmatrix} 1 & 1 & 0 & \beta & -2 & \beta \\ 0 & \beta & -2 & \beta & -2 & -\beta \\ 0 & 1 & \alpha & 1 & \alpha & \alpha\beta+1 \end{pmatrix},$$

which clearly has full rank for all α, β. Hence, the given system is controllable for all α, β.

(b) When only the second control acts, that is to say, when $u_1(t) = 0$ for all t, the system can be written

$$\dot{x}(t) = Ax(t) + b_2 u_2(t).$$

Then, the controllability matrix is reduced to

$$K(A,b_2) = \begin{pmatrix} 1 & \beta & \beta \\ \beta & \beta & -\beta \\ 1 & 1 & \alpha\beta+1 \end{pmatrix},$$

which has full rank for $\beta \neq 0$, $\alpha \neq \frac{-2}{\beta}$.

(c) Now one has

$$K(A,B) = \left(\begin{array}{ccccc|ccc|cc} 0 & 0 & 0 & 0 & 0 & \gamma & 0 & \delta & 1 & 0 \\ 0 & 0 & 0 & \gamma & 0 & \delta & 1 & 0 & 0 & 0 \\ 0 & \gamma & 0 & \delta & 1 & 0 & 0 & 0 & 0 & 0 & 0 & 0 \\ 0 & \delta & 1 & 0 & 0 & 0 & 0 & 0 & 0 & 0 \\ 1 & 0 & 0 & 0 & 0 & 0 & 0 & 0 & 0 & 0 \\ \hline 0 & 0 & 1 & 0 & 0 & 0 & 0 & 0 \\ 0 & 1 & 0 & 0 & 0 & 0 & 0 & 0 & 0 \end{array} \right).$$

The "controllability indices" can be computed from the ranks of the $2, 4, 6, \ldots$ first columns:

$$\text{rank}(B) = 2, \text{ for all } \gamma, \delta$$
$$\text{rank}(B, AB) = 4, \text{ for all } \gamma, \delta$$
$$\text{rank}(B, AB, A^2B) = 5 \text{ if } \gamma = \delta = 0; \quad \text{rank}(B, AB, A^2B) = 6 \text{ otherwise}$$
$$\text{rank}(B, \ldots, A^3B) = 6 \text{ if } \gamma = \delta = 0; \quad \text{rank}(B, \ldots, A^3B) = 7 \text{ otherwise}$$
$$\text{rank}(B, \ldots, A^4B) = 7$$

3. Solution

(1) Let

$$G(s) = \begin{pmatrix} 1/s \\ 1/(s-1) \end{pmatrix}.$$

Then: $p = 2, m = 1$. Following the given pattern:

(i) $P(s) = s(s-1) = -s + s^2$
$r = 2, p_0 = 0, p_1 = -1, p_2 = 1$

(ii) $G(s) = \frac{1}{-s+s^2} \begin{pmatrix} s-1 \\ s \end{pmatrix}$

(iii) $G(s) = \frac{1}{-s+s^2} \left(\begin{pmatrix} -1 \\ 0 \end{pmatrix} + \begin{pmatrix} 1 \\ 1 \end{pmatrix} s \right)$

$R_0 = \begin{pmatrix} -1 \\ 0 \end{pmatrix}, R_1 = \begin{pmatrix} 1 \\ 1 \end{pmatrix}$

(iv) $A = \begin{pmatrix} 0 & 1 \\ 0 & 1 \end{pmatrix}, \quad B = \begin{pmatrix} 0 \\ 1 \end{pmatrix}, \quad C = \begin{pmatrix} -1 & 1 \\ 0 & 1 \end{pmatrix}$

(2) $\text{rank} K(A, B) = \text{rank} \begin{pmatrix} 0 & 1 \\ 1 & 1 \end{pmatrix} = 2$

(3) $(sI - A)^{-1} = \left(s \begin{pmatrix} 1 & 0 \\ 0 & 1 \end{pmatrix} - \begin{pmatrix} 0 & 1 \\ 0 & 1 \end{pmatrix} \right)^{-1} = \begin{pmatrix} s & -1 \\ 0 & s-1 \end{pmatrix}^{-1} = \frac{1}{s^2-s} \begin{pmatrix} s-1 & 1 \\ 0 & s \end{pmatrix}$

$C(sI - A)^{-1}B = \begin{pmatrix} -1 & 1 \\ 0 & 1 \end{pmatrix} \frac{1}{s^2-s} \begin{pmatrix} s-1 & 1 \\ 0 & s \end{pmatrix} \begin{pmatrix} 0 \\ 1 \end{pmatrix} = \frac{1}{s^2-s} \begin{pmatrix} -1 & 1 \\ 0 & 1 \end{pmatrix} \begin{pmatrix} 1 \\ s \end{pmatrix}$

$= \frac{1}{s^2-s} \begin{pmatrix} -1+s \\ s \end{pmatrix} = G(s)$

4. Solution

(1) $x = (-1, 2, 1)$ is 3-step reachable, from the origin, if there are $u_2(0), u_2(1), u_2(2)$ such that:

$$x = A^2 b_1 u_1(0) + A b_1 u_1(1) + b_1 u_1(2)$$

$$= \begin{pmatrix} 10 \\ -8 \\ 8 \end{pmatrix} u_1(0) + \begin{pmatrix} 2 \\ -2 \\ 2 \end{pmatrix} u_1(1) + \begin{pmatrix} 0 \\ -1 \\ 1 \end{pmatrix} u_1(2)$$

or equivalently::

$$\begin{pmatrix} -1 \\ 2 \\ 1 \end{pmatrix} = \begin{pmatrix} 10 & -2 & 0 \\ -8 & -2 & -1 \\ 8 & 2 & 1 \end{pmatrix} \begin{pmatrix} u_1(0) \\ u_1(1) \\ u_1(2) \end{pmatrix}$$

No solutions exist because

$$\text{rank}\begin{pmatrix} 10 & -2 & 0 \\ -8 & -2 & -1 \\ 8 & 2 & 1 \end{pmatrix} = 2 \qquad \text{rank}\begin{pmatrix} 10 & -2 & 0 & -1 \\ -8 & -2 & -1 & 2 \\ 8 & 2 & 1 & 1 \end{pmatrix} = 3$$

(2) Analogously:

$$\begin{pmatrix} -1 \\ 2 \\ 1 \end{pmatrix} = \begin{pmatrix} -1 & 0 \\ 1 & 1 \\ 1 & 0 \end{pmatrix} \begin{pmatrix} u_2(0) \\ u_2(1) \end{pmatrix}$$

whose solution is: $u_2(0) = 1$, $u_2(1) = 1$.

(3.1) Analogously:

$$\begin{pmatrix} -1 \\ 2 \\ 1 \end{pmatrix} = \begin{pmatrix} 2 & -1 & 0 & 0 \\ -2 & 1 & -1 & 1 \\ 2 & 1 & 1 & 0 \end{pmatrix} \begin{pmatrix} u_2(0) \\ u_2(0) \\ u_1(1) \\ u_2(1) \end{pmatrix}$$

The solutions can be parameterized by $u_1(0)$ as follows:

$$u_2(0) = 1 + u_1(0)$$
$$u_1(1) = 1 - 2u_1(0) - u_2(0) = -4u_1(0)$$
$$u_2(1) = \ldots = 1 - 4u_1(0)$$

(3.2) It is not possible, because $u_1(0)$ and $u_1(1)$ have opposite signs.

(3.3) It is possible $u_1(0) = u_2(0) = -1$. Then $u_1(1) = 4$, $u_2(1) = 5$.

5. Solution

(1.1) When $x(0) = 0$, one has:

$$x(k) = (A^{k-1}b_1u_1(0) + \ldots + A^{k-1}b_mu_m(0))$$

$$\vdots$$

$$+ (Ab_1u_1(k-2) + \ldots + Ab_mu_m(k-2))$$
$$+ (b_1u_1(k-1) + \ldots + b_mu_m(k-1))$$

where $B = (b_1, \ldots, b_m)$ and $u(0), \ldots, u(k-1)$ run over all possible control functions. Therefore

$$K(h) = [A^{k-1}b_1, \ldots, A^{k-1}b_m, \ldots, Ab_1, \ldots, Ab_m, b_1, \ldots, b_m] = [B, AB, \ldots, A^{k-1}B].$$

(1.2) It is obvious from (1.1).

(1.3) $K(h) = K(h+1)$ if and only if $A^h b_1, \ldots, A^h b_m \in K(h)$.

Then $A^{h+1} b_1, \ldots, A^{h+1} b_m \in K(h+1)$, so that $K(h+2) = K(h+1)$.

(1.4) As $\dim K(h) \le n$, the length of the increasing chain is n at most.

(2)

$$K_i + K_j = [b_i, Ab_i, \ldots, A^{n-1} b_i] + [b_j, Ab_j, \ldots, A^{n-1} b_j]$$
$$= [(b_i, b_j), A(b_i, b_j), \ldots, A^{n-1}(b_i, b_j)]$$

(3.1) $K = \mathrm{Im} \begin{pmatrix} 0 & 0 & 2 & -1 & 10 & -2 \\ -1 & 1 & -2 & 1 & -8 & 2 \\ 1 & 0 & 2 & 1 & 8 & 2 \end{pmatrix} = \mathbb{R}^3$

$K_1 = \mathrm{Im} \begin{pmatrix} 0 & 2 & 10 \\ -1 & -2 & -8 \\ 1 & 2 & 8 \end{pmatrix} = \mathrm{Im} \begin{pmatrix} 0 & 2 \\ -1 & -2 \\ 1 & 2 \end{pmatrix}$

A basis of K_1: $(0, -1, 1)$, $(1, 0, 0)$

$K_2 = \mathrm{Im} \begin{pmatrix} 0 & -1 & -2 \\ 1 & 1 & 2 \\ 0 & 1 & 2 \end{pmatrix} = \mathrm{Im} \begin{pmatrix} 0 & -1 \\ 1 & 1 \\ 0 & 1 \end{pmatrix}$

A basis of K_2: $(0, 1, 0)$, $(-1, 0, 1)$

(3.2) $K_1 + K_2 = \mathrm{Im} \begin{pmatrix} 0 & 1 & 0 & -1 \\ -1 & 0 & 1 & 0 \\ 1 & 0 & 0 & 1 \end{pmatrix} = \mathbb{R}^3$

$K_1 = \{y + z = 0\}$
$K_2 = \{x + z = 0\}$
$K_1 \cap K_2 = \{y + z = x + z = 0\}$
A basis of $K_1 \cap K_2$: $(1, 1, -1)$

6. Solution

(1) If $\bar{x} = S^{-1} x$, then:

$$\dot{\bar{x}} = S^{-1} \dot{x} = S^{-1}(Ax + Bu) = S^{-1} A S \bar{x} + S^{-1} B u$$
$$y = Cx = CS\bar{x}$$

(2) $\mathrm{rank}(\bar{B}, \bar{A}\bar{B}, \ldots, \bar{A}^h \bar{B}) = \mathrm{rank}(S^{-1}B, S^{-1}ASS^{-1}B, \ldots, (S^{-1}AS)^h S^{-1}B)$
$= \mathrm{rank}\, S^{-1}(B, AB, \ldots, A^h B) = \mathrm{rank}(B, AB, \ldots, A^h B)$

(3) $\bar{G}(s) = \bar{C}(sI - \bar{A})^{-1} \bar{B} = CS(sI - S^{-1}AS)^{-1}S^{-1}B = CS(S^{-1}(sI - A)S)^{-1}S^{-1}B$
$= CSS^{-1}(sI - A)^{-1}SS^{-1}B = G(s)$

7. Solution

(1) From Cayley-Hamilton theorem:

$$A^n = a_0 I + a_1 A + \ldots + a_{n-1} A^{n-1}$$

Therefore, if $x \in \text{Im}\left(B \; AB \; \ldots \; A^{n-1}B\right)$, then

$$Ax \in \text{Im}\left(AB \; A^2B \; \ldots \; A^nB\right) \subset \text{Im}\left(B \; AB \; \ldots \; A^nB\right),$$

and

$$\text{Im}\left(B \; AB \; \ldots \; A^nB\right) = \text{Im}\left(B \; AB \; \ldots \; (a_0B + a_1AB + \ldots + a_{n-1}A^{n-1}B)\right)$$

$$= \text{Im}\left(B \; AB \; \ldots \; A^{n-1}B\right)$$

$$= K.$$

If $x \in L$, then:

$$Cx = CAx = \ldots = CA^{n-1}x = 0$$

Clearly:

$$C(Ax) = \ldots = CA^{n-2}(Ax) = 0$$

It is sufficient to prove that $CA^{n-1}(Ax) = 0$, but

$$CA^{n-1}(Ax) = CA^nx$$

$$= C(a_0I + a_1A + \ldots + a_{n-1}A^{n-1})x$$

$$= a_0Cx + a_1CAx + \ldots + a_{n-1}CA^{n-1}x = 0$$

(2.1) $K = \text{Im} \begin{pmatrix} 1 & 1 & 1 & 1 \\ 0 & 0 & 0 & 0 \\ 0 & 1 & -2 & 3 \\ 1 & -1 & 1 & -1 \end{pmatrix}$; $\dim K = 3$

$L = \text{Ker} \begin{pmatrix} 0 & 1 & 0 & 1 \\ 0 & 1 & 0 & -1 \\ 0 & 1 & 0 & 1 \\ 0 & 1 & 0 & -1 \end{pmatrix}$; $\dim L = 4 - 2 = 2$

(2.2) Basis of K: (u_1, u_2, u_3)
$u_1 = (1,0,0,1)$, $u_2 = (1,0,1,-1)$, $u_3 = (1,0,-2,1)$
Basis of L: (v_1, v_2)
$v_1 = (1,0,0,0)$, $v_2 = (0,0,1,0)$

(2.3) $Au_1 = u_2$; $Au_2 = u_3$; $Au_3 = \begin{pmatrix} 1 \\ 0 \\ 3 \\ -1 \end{pmatrix} = u_1 + u_2 - u_3$

$\text{Mat } A|_K = \begin{pmatrix} 0 & 0 & 1 \\ 1 & 0 & 1 \\ 0 & 1 & -1 \end{pmatrix}$

$Av_1 = v_1$; $Av_2 = -v_2$

$\text{Mat } A|_L = \begin{pmatrix} 1 & 0 \\ 0 & -1 \end{pmatrix}$

(2.4) $K = \{(x_1, x_2, x_3, x_4) : x_2 = 0\}$
$L = \{(x_1, x_2, x_3, x_4) : x_2 = x_4 = 0\}$
$K \cap L = L$

8. Solution

(1) We recall (see ex. 7) that K and L are A-invariant subspaces. Therefore:

$A(S_2) \subset [S_2]$

$A(S_1) \subset [S_1, S_2]$

$A(S_4) \subset [S_2, S_4]$

so that $\bar{A} = S^{-1}AS$ has the stated form.
Moreover $\operatorname{Im} B \subset K$. Hence, $\bar{B} = S^{-1}B \subset [S_1, S_2]$.
Finally, as $L \subset \operatorname{Ker} C$, we have $C(S_2) = C(S_4) = 0$.

(2.1) According to the solution of ex. 7, we can take:

$$S_2 = \begin{pmatrix} 1 & 0 \\ 0 & 0 \\ 0 & 1 \\ 0 & 0 \end{pmatrix} \quad S_1 = \begin{pmatrix} 0 \\ 0 \\ 0 \\ 1 \end{pmatrix} \quad S_3 = \begin{pmatrix} 0 \\ 1 \\ 0 \\ 0 \end{pmatrix}$$

and S_4 is empty. Hence:

$$S = \begin{pmatrix} 0 & 1 & 0 & 0 \\ 0 & 0 & 0 & 1 \\ 0 & 0 & 1 & 0 \\ 1 & 0 & 0 & 0 \end{pmatrix}$$

(2.2)

$$\bar{A} = S^{-1}AS = \begin{pmatrix} -1 & 0 & 0 & 0 \\ 0 & 1 & 0 & 1 \\ 1 & 0 & -1 & 0 \\ 0 & 0 & 0 & 1 \end{pmatrix}$$

$$\bar{B} = S^{-1}B = \begin{pmatrix} 1 \\ 1 \\ 0 \\ 0 \end{pmatrix} \quad \bar{C} = CS = (1 \mid 0\ 0 \mid 1)$$

That is to say:

$$A_{11} = (-1) \quad A_{13} = (0\ 0) \quad A_{21} = \begin{pmatrix} 0 \\ 1 \end{pmatrix} \quad A_{22} = \begin{pmatrix} 1 & 0 \\ 0 & -1 \end{pmatrix} \quad A_{23} = \begin{pmatrix} 1 \\ 0 \end{pmatrix}$$

$$A_{33} = (1) \quad B_1 = (1) \quad B_2 = \begin{pmatrix} 1 \\ 0 \end{pmatrix} \quad C_1 = (1) \quad C_3 = (1)$$

(2.3) $\operatorname{rank} K(A_{11}, B_1) = \operatorname{rank}(1\ -1) = 1$
$\operatorname{rank}(L(C_1, A_{11}))^t = \operatorname{rank}(1\ -1) = 1$

$$\text{rank } K\left(\begin{pmatrix} A_{11} & 0 \\ A_{21} & A_{22} \end{pmatrix}, \begin{pmatrix} B_1 \\ B_2 \end{pmatrix}\right) = \text{rank}\begin{pmatrix} 1 & -1 & 1 \\ 1 & 1 & 1 \\ 0 & 1 & -2 \end{pmatrix} = 3$$

$$\text{rank }\left(L\left((C_1\ C_3), \begin{pmatrix} A_{11} & 0 \\ A_{21} & A_{33} \end{pmatrix}\right)\right)^t = \text{rank}\begin{pmatrix} 1 & -1 \\ 1 & 1 \end{pmatrix} = 2$$

9. Solution

(1.1) It is easy to check that there is only one main minor of each sign. Then:

$$a_1 = \text{tr } A_c, \quad a_2 = -\det\begin{pmatrix} 0 & 1 \\ a_2 & a_1 \end{pmatrix}, \quad a_3 = \det\begin{pmatrix} 0 & 1 & 0 \\ 0 & 0 & 1 \\ a_3 & a_2 & a_1 \end{pmatrix}, \dots, \quad a_n = \det A_c.$$

(1.2) If $F_c = (f_1 \dots f_n)$, then

$$A_c + b_c F_c = \begin{pmatrix} 0 & 1 & \cdots & 0 & 0 \\ 0 & 0 & \cdots & 0 & 0 \\ & & \cdots & & \\ 0 & 0 & \cdots & 0 & 1 \\ a_n + f_1 & a_{n-1} + f_2 & \cdots & a_2 + f_{n-1} & a_1 + f_n \end{pmatrix}$$

Therefore, $f_1, \dots, f_n$ are the solutions of:

$$t^n - (a_1 + f_n)t^{n-1} - \dots - (a_{n-1} + f_2)t - (a_n + f_1) = (t - \mu_1)(t - \mu_2) \cdots (t - \mu_n).$$

(1.3)

$$\text{eigenvalues}(A + bF) = \text{eigenvalues}(SA_cS^{-1} + Sb_cF_cS^{-1})$$
$$= \text{eigenvalues}(S(A_c + b_cF_c)S^{-1})$$
$$= \text{eigenvalues}(A_c + b_cF_c) = \{\mu_1, \dots, \mu_n\}.$$

(2.1) It is straightforward to check that:

$$S^{-1}AS = \begin{pmatrix} 0 & 1 & 0 \\ 0 & 0 & 1 \\ -1 & 2 & 0 \end{pmatrix}, \quad S^{-1}b = \begin{pmatrix} 0 \\ 0 \\ 1 \end{pmatrix}.$$

(2.2) First, we look for $F_c = (f_1\ f_2\ f_3)$ such that:

$$t - f_3t^2 - (2 + f_2) - (-1 + f_1) = (t - \mu_1)(t - \mu_2)(t - \mu_3)$$

Therefore:

$$f_3 = \mu_1 + \mu_2 + \mu_3$$
$$f_2 = \mu_1\mu_2 + \mu_2\mu_3 + \mu_3\mu_1 - 2$$
$$f_1 = \mu_1\mu_2\mu_3 + 1$$

Finally: $F = F_c S^{-1}$.

10. Solution

(1) From Cayley-Hamilton theorem, if the characteristic polynomial of A is

$$Q(t) = (-1)^n (t^n - a_1 t^{n-1} - \ldots - a_{n-1} t - a_n),$$

then

$$A^n = a_n I + a_{n-1} A + \ldots + a_1 A^{n-1}.$$

Therefore, if $x \in K = \mathrm{Im} \left(B \ AB \ \ldots \ A^{n-1} B \right)$, then

$$Ax \in \mathrm{Im} \left(AB \ A^2 B \ \ldots \ A^n B \right)$$

$$\subset \mathrm{Im} \left(B \ AB \ \ldots \ A^{n-1} B \ (a_0 B + \ldots + a_{n-1} A^{n-1} B) \right) = K$$

(2) If $(u_1, \ldots, u_d, \ldots, u_m)$ is a basis of the state space adapted to K, then a state x belongs to K if and only if the last $n - d$ coordinates are 0.

(3) In the conditions of (2), the d first columns of $\bar{A}$ are $Au_1, \ldots, Au_d$, which belong to K (see (1)). Hence, again from (2), their last $n - d$ coordinates are 0. The same argument works for B, because its columns belong to K.

(4) From (3), it is clear that $x_{uc}(0) = 0$ implies $x_{uc}(t) = 0$ for any control $u(t)$.

(5) Again from (3), if $x_{uc}(t) = 0$, the remainder coordinates $x_c(t)$ are determined by system (2).

(6) By hypothesis

$$\mathrm{rank} \left(B \ AB \ \ldots \ A^{n-1} B \right) = d.$$

The rank is preserved under changes of bases. Hence:

$$d = \mathrm{rank} \left(\bar{B} \ \bar{A}\bar{B} \ \ldots \ \bar{A}^{n-1}\bar{B} \right) = \mathrm{rank} \left(\begin{pmatrix} B_c \\ 0 \end{pmatrix} \begin{pmatrix} A_c B_c \\ 0 \end{pmatrix} \ldots \begin{pmatrix} A_c^{n-1} B_c \\ 0 \end{pmatrix} \right)$$

$$= \mathrm{rank} \left(B_c \ A_c B_c \ \ldots \ A_c^{n-1} B_c \right) = \mathrm{rank} \left(B_c \ A_c B_c \ \ldots \ A_c^{d-1} B_c \right)$$

where the last equality follows from Cayley-Hamilton theorem.

(7) Recall that the transfer matrix is preserved under changes of bases:

$$C(sI - A)^{-1} B = \bar{C} S^{-1} (sI - S\bar{A}S^{-1})^{-1} S\bar{B} = \bar{C} S^{-1} (S(sI - \bar{A})S^{-1})^{-1} S\bar{B} =$$

$$= \bar{C} S^{-1} (S^{-1})^{-1} (sI - \bar{A}) S^{-1} S\bar{B} = \bar{C}(sI - \bar{A})\bar{B}$$

From (2):

$$(sI - \bar{A})^{-1} = \left(sI - \begin{pmatrix} A_c & * \\ * & * \end{pmatrix} \right)^{-1} = \begin{pmatrix} sI_d - A_c & * \\ 0 & * \end{pmatrix}^{-1} = \begin{pmatrix} (sI_d - A_c)^{-1} & * \\ 0 & * \end{pmatrix}$$

where it does not matter the form of the blocks *:

$$\bar{C}(sI - \bar{A})^{-1}\bar{B} = \begin{pmatrix} C_c & C_{uc} \end{pmatrix} \begin{pmatrix} (sI_d - A_c)^{-1} & * \\ 0 & * \end{pmatrix} \begin{pmatrix} B_c \\ 0 \end{pmatrix}$$

$$= \begin{pmatrix} C_c & C_{uc} \end{pmatrix} \begin{pmatrix} (sI_d - A_c)^{-1}B_c \\ 0 \end{pmatrix}$$

$$= C_c(sI_d - A_c)^{-1}B_c$$

11. Solution

(1)

$$A = \begin{pmatrix} 0 & 0 & -1 \\ 0 & 1 & -1 \\ 1 & 0 & 1 \end{pmatrix}; \quad b = \begin{pmatrix} 1 \\ 1 \\ 1 \end{pmatrix}$$

(1.1)

$$\text{rank}\begin{pmatrix} b & Ab & A^2b \end{pmatrix} = \text{rank}\begin{pmatrix} 1 & 1 & 2 \\ 1 & 0 & -2 \\ 1 & 2 & 3 \end{pmatrix} = 3$$

(1.2)

$$S_1 = \begin{pmatrix} 2 & 1 & 1 \\ -2 & 0 & 1 \\ 3 & 2 & 1 \end{pmatrix}; \quad S_1^{-1} = \frac{1}{3}\begin{pmatrix} 2 & -1 & -1 \\ -5 & 1 & 4 \\ 4 & 1 & -2 \end{pmatrix}$$

$$\bar{A} = \frac{1}{3}\begin{pmatrix} 6 & 3 & 0 \\ 0 & 0 & 3 \\ -3 & 0 & 0 \end{pmatrix} \quad \bar{b} = \frac{1}{3}\begin{pmatrix} 2 & -1 & -1 \\ -5 & 1 & 4 \\ 4 & 1 & -2 \end{pmatrix}\begin{pmatrix} 1 \\ 1 \\ 1 \end{pmatrix} = \frac{1}{3}\begin{pmatrix} 0 \\ 0 \\ 3 \end{pmatrix}$$

(1.3)

$$S_2 = \begin{pmatrix} 1 & 0 & 0 \\ -2 & 1 & 0 \\ 0 & -2 & 1 \end{pmatrix}; \quad S_2^{-1} = \begin{pmatrix} 1 & 0 & 0 \\ 2 & 1 & 0 \\ 4 & 2 & 1 \end{pmatrix}$$

$$A_c = \begin{pmatrix} 0 & 1 & 0 \\ 0 & 0 & 1 \\ -1 & 0 & 2 \end{pmatrix}; \quad b_c = \begin{pmatrix} 1 & 0 & 0 \\ 2 & 1 & 0 \\ 4 & 2 & 1 \end{pmatrix}\begin{pmatrix} 0 \\ 0 \\ 1 \end{pmatrix} = \begin{pmatrix} 0 \\ 0 \\ 1 \end{pmatrix}$$

(2) The computations are analogous to those in (1).

(3.1) If λ_i is an eigenvalue of A_c, then:

$$\text{rank}(A_c - \lambda_i I) < n.$$

On the other hand,

$$\text{rank}(A_c - \lambda_i I) = \text{rank}\begin{pmatrix} -\lambda_i & 1 & 0 & \dots & 0 & 0 \\ 0 & -\lambda_i & 1 & \dots & 0 & 0 \\ & & \dots & \dots & & \\ 0 & 0 & 0 & \dots & -\lambda_i & 1 \\ * & * & * & \dots & * & * \end{pmatrix} \geq n-1.$$

(3.2) If λ_i is an eigenvalue of A, then λ_i is also en eigenvalue of A_c. Then, from (3.1):

$$\dim \mathrm{Ker}(A - \lambda_i I) = n - \mathrm{rank}(A - \lambda_i I) = n - \mathrm{rank}(A_c - \lambda_i I) = n - (n-1) = 1.$$

Therefore, A is non-diagonalizable if the algebraic multiplicity of λ_i is greater than 1.

4. References

Chen, Chi-Tsong (1984). *Linear System Theory and Design*, Holt-Saunders International Editions, Japan. ISBN: 4-8337-0191-X.

Ferrer, J.; Ortiz-Caraballo, C. & Peña, M. (2010). Learning engineering to teach mathematics, *40th ASEE/IEEE Frontiers in Education Conference*

Kalman, R. E.; Falb, P. L. & Arbib, M. A. (1974). *Topics in mathematical system theory*, PTMH, India, McGraw-Hill

Lay, D.C (2007). *Algebra lineal y sus aplicaciones*, Ed. Pearson, Mexico. ISBN: 978-970-26-0906-3

Puerta Sales, F. (1976). *Algebra Lineal*, Universidad Politecnica de Barcelona. ETS Ingenieros Industriales de Barcelona in collaboration with Marcombo Boixareu Ed. ISBN: 84-600-6834-X

9

Introduction to the Computer Modeling of the Plague Epizootic Process

Vladimir Dubyanskiy[1], Leonid Burdelov[2] and J. L. Barkley[3]
[1]Stavropol Plague Control Research Institute,
[2]M. Aikimbaev's Kazakh Science Center of Quarantine and Zoonotic Diseases,
[3]Virginia Polytechnic Institute and State University
[1]Russian Federation
[2]Republic of Kazakhstan
[3]USA

1. Introduction

Bubonic plague, caused by the bacteria, Yersinia pestis, persists as a public health problem in many parts of the world, including Central Asia and Kazakhstan. Plague is a vector-borne disease, i.e. the disease is spread by arthropod vectors that live and feed on hosts (Gage and Kosoy, 2005). Great gerbil (Rhombomys opimus Liht., 1823, Rodentia, Cricetidae) is a significant plague host in Central Asia and Kazakhstan region. Transmission happens through bites of infected fleas that earlier fed on infected hosts. This plague epizootic process study can help future plague control work.

Modeling the epizootic process of plague in great gerbil settlements allows for quantitative analyses of epizootic characteristics, which we are unable to control for in nature (Soldatkin et al, 1973). Only one such attempt of this type of modeling is known (Soldatkin et al, 1966; Soldatkin and Rudenchik, 1967; Soldatkin et al, 1973). In this study, the authors demonstrated the high potential for this model, yet it possessed many limitations and had a small work space (4,096 holes). The development of high-performance workstations, and integration of geographic information systems (GIS) and remote sensing now makes it possible to model complex, epizootic processes of plague.

We created a probabilistic, cellular automaton model of the epizootic process of plague by using all of the above-named technologies. A probabilistic, cellular automaton with Monte-Carlo, as the chosen statistical method, is the basic mechanism of the model (Etkins, 1987; Grabovskiy, 1995). It is the SEIR (susceptible, exposed, infected and removed) model with specific additions.

2. Description of the model

The model was created using Microsoft Excel 2003 with Visual Basic as the code language. Microsoft Excel has 255 columns and 65,000 rows. A user-defined, fixed size of the modeling area depends on the density of the gerbil colonies. A probabilistic, cellular automaton with Monte-Carlo, as the chosen statistical method, is the basic mechanism of the model

(Dubyanskiy V.M., Burdelov L.A. , 2010, Dubyanskiy V.M, 2010). The choice for this model type was rationalized by the following:

- Commonly-used mathematical models that rely on differential equations require input from many data sources. While, at the same time, the availability of plague surveillance source data is rather insufficient.
- Temporal models, (based on differential equations or other methods), are being used for modeling without consideration of spatial distribution. In these models, it is assumed that all events are happening at an abstract point rather than considering plague focus, landscape region, or primary square.
- If plague simulates nature, it analyzes only a static structure of plague focus or its parts.

The probabilistic, cellular automaton simulation allows the user to combine all of these rationales. The epizootic process of plague in Great gerbil settlements was a good subject for this experiment. The gerbil's burrow system represents a time-spatial, discrete unit of an epizootic process (Sedin, 1985). The data on burrow systems distribution within a plague focus was determined by GIS and remote sensing. Satellite images of the burrow systems provide a coherent picture of the colony structure (Burdelov et al., 2007; Addink et al., 2010) at many plague foci (Fig 1.).

The cell/row structure of the model's workspace allowed us to observe the proportion between burrow systems and settlements and density of colonies. The density of colonies was the main factor approximating the number of rodents. Both the density of colonies and the number of rodents are significant factors for plague prediction (Klassovskiy et al., 1978; Davis et al., 2004, 2008).

Several quantitative and qualitative relationships and characteristics between components of the plague's parasitic system were identified early on (Burdelov et al., 1984; Begon et al., 2006). Analyzing these numerous, character-combinations required that we use Monte-Carlo simulation, rather than differential equation models, to describe each specific time stamp in question. The analysis of burrow systems distribution on satellite images demonstrated that the number of nearest neighbors in the burrow system often exceeded four (von Neumann neighborhood). Hence, our model used the square (eight-cell) Moore neighborhood (Fig 2.).

The state of colonies can be categorized as "infected", "infecting", "immune", or "readied" for reinfecting. An infected colony is one that is infected, (in host, in vector or in substratum), but has not spread the plague microbe (latent period). An infecting colony is one that is infected and is spreading the plague microbe among other colonies. A colony that has been cleared of the plague microbe and is not infectious is called an immune colony. A readied colony is one which can be infected again after the immune period.

Data on the vector's and host's infection were used to the extent that it was necessary for calculating the model's parameters. An infecting agent can be transmitted to other colonies independent of the main mechanism of "host-vector-host" transmission. These processes define whether a colony is infected or not and constitutes the "black box" of the model. Nuances of an epizootic process which might be connected to specific mechanisms of plague transmission have not been reviewed.

The cycle of infection is a duration when an infected colony is being converted to an infecting colony and can spread the plague microbe among other colonies. The obligatory scan of the model's workspace occurs at the end of each cycle of infection. The probability of transfer success is defined as one attempt of the plague microbe to transfer from an infected colony to a non-infected colony. The number of attempts required to infect neighboring colonies from one "infecting" colony during one cycle of infection is termed the coefficient of colony infection. This term was proposed by Litvin et al., (1980). However, this term, in our case, unites the coefficient of infection of a colony by Litvin and the coefficient of transmission by Soldatkin et al., (1973). The rate (or the value) of epizootic contact is the product of the successive transfer probability and the coefficient of a colony infection.

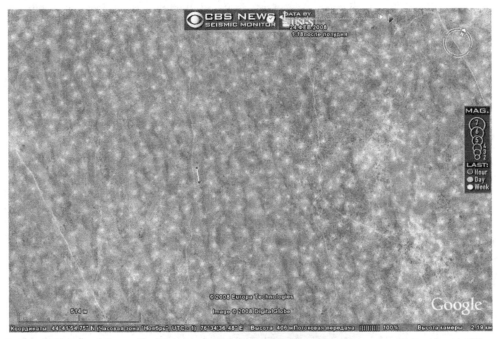

Fig. 1. Colonies of Great Gerbil in a satellite image. Each bright disc represents colonies 10-40m in diameter. The image was captured using the publicly available software Google Earth (http://earth.google.com/). Copyright 2008 DigitalGlobe; Europa Technologies.

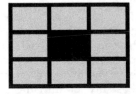

Fig. 2. Moore's environment allows using all cells in model for its greater efficiency.

$$L = C_i * C_t \tag{1}$$

where L is a rate (or the value) of epizootic contact, C_i the coefficient of colony infection, C_t the probability of transfer success.

The relative remote drift is the possibility of more or less remote colonies infecting without nearer colonies infecting. This parameter can be regulated by the change of plague drift distance as well as by the change of plague drift realization probability. It also allows changing the speed of epizootic spread.

The structure of great gerbil settlements was reflected by altering colony-cells and non colony-cells. This allowed working with different configurations of settlements and with different densities of burrow systems. The scheme of burrow system's spatial location could be integrated into the work space of the model from a map or from a satellite image if necessary (Fig. 3). If the epizootic process was imitated, then the spatial location of burrow systems would be defined by a random-number generator. Initial results were published earlier (Dubyanskiy, Burdelov, 2008).

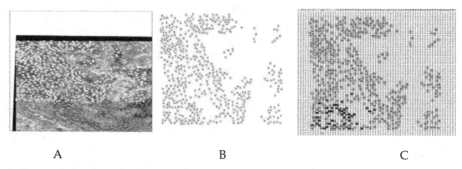

A   B   C

Fig. 3. Carry of a real configuration of settlement of Great Gerbil from the satellite image adhered in GIS in working space of model. A - Allocated by means of GIS colonies of Great Gerbil. B - Layer of GIS ready for carry to working space of model. C - Working space of model with developing of plague epizooty.

There are two ways of spreading the plague microbe. The "infected" colony could attempt to infect the other colony-cell as well as non colony-cell. In the latter case, the attempt of infecting is useless. However, the tendency of gerbils to migrate among colonies will spread the plague microbe to occupied colonies. The probability of transfer success is a probability that a sick flea (or sick gerbil) will be transferred from an "infected" colony to a non-infected colony and a microbiocenose of colony will be involved in the epizootic process. The coefficient of a colony infection demonstrates how often transfers occurred (i.e. a gerbil can visit non-infected colony two times or 10 times and so on).

The basic quantitative and qualitative characteristics were entered into the parameter sheet. The value of parameters can be changed before modeling, although the dynamic changing of parameters is possible. The parameters could be, for example, quantities of the "cycle of infection" for one season (or session of modeling), the duration of the cycle of infection, the density of colonies, and the probability of infecting neighbor colonies. The technical conditions of the model work were also defined. The built-in menu allows comfortably managing the modeling options.

Cellular flattened models have the edge effect. Naturally, this effect took place in our model too. It was decided that all transmission of plague microbes outside the model workspace was not consequential. In this case the model was made simpler. The model settlement looked like an isolated island of the gerbil's settlement in nature.

2.1 Basic parameters of the model

Even with the seeming simplicity of the model, the real quantity of fulfilled interrelations is rather great. For example, "infection" of a colony depends on: the duration of cycle of infection; the duration of the colony's status as infected; infectious quotient and immunity; the probability of success of transfer; the coefficient of infection of a colony; the density and distribution of colonies in the workspace of the model; and the number and arrangement of infected colonies. Considering that further behavior of a cell depends on probability and status of surrounding cells at PCA, to give the optimum and full analysis of influence of such quantities of the interconnected parameters is very difficult. Therefore, initial selection of the basic working parameters of the model were carried out in a literature search of data, including results of direct or physical modeling of epizootic process using radioactive isotopes in real settlements of the great gerbils. It is necessary to make a stipulation that the analysis of literature has shown an absence of any unequivocal quantitative characteristic of epizootic process in a required aspect.

After infection by the plague microbe, a colony becomes "infected". This period in the model lasts for 10 days where three days are the time of bacteremia and seven days are the time of flea's prohibition (block of the proventriculus) if the flea fed on a sick animal. The colony has the possibility to infect neighbors from day 11 to 50 from the beginning of its own infection. It is not infectious from day 51 to 90, and it is immune (Novikova et al., 1971; Naumov et al., 1972). The colony can be infected again after 90 days (Samsonovich et al., 1971; Rothschild et al., 1975; Rothschild, 1978).

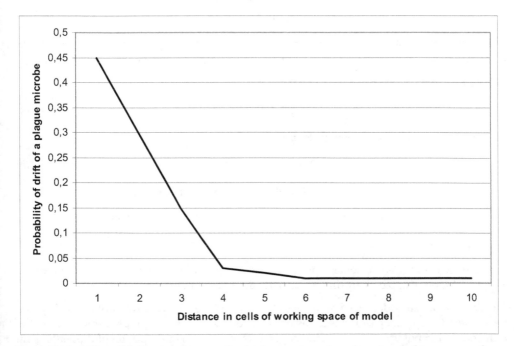

Fig. 4. Probability of drift of a plague microbe from colony to colony in working space of model during one cycle of infection.

The probabilities distribution of plague microbe drift approximately corresponds to data of direct modeling of drift of the fleas in the neighboring holes (Soldatkin, Rudenchik, 1971), and to data on movement distances of gerbils to the neighbor colonies (Alekseev, 1974; Davis et al., 2008) (Fig. 4). The probability of transfer success could be modified by the user within the limits from zero to one. The coefficient of a colony infection could be varied from one up to experimentally received or any other logically admissible value. The density, distribution, occupation, and number of a plague microbe hosts expressed through occupation were given as cell-colonies percentage, from quantity of cells in all working spaces of the model and could be changed from zero to 100. It would be desirable to emphasize that it was possible to change any parameters in the model.

The maximum drift of plague microbe in model can mark as l. Every infecting colony can infect the neighbours in a square with diagonal $2l$. The infecting colony is situated in the center of this square. The probability of contact with any colonies inside this square defined as

$$Pz = \frac{N_l - (i_1 + i_2 + R)}{N_l} \tag{2}$$

where Pz is the probability of contact with the neighbour colony, N_l is the general number of colonies in the square with $2l$ diagonal, i_1 is the infected colony, i_2 is the infecting colony, R is the immune colony.

The probability of colony infect is defined as

$$Pi = Pz * Ct \; Ci \tag{3}$$

where Pi is the probability of colony infect, Ct is the probability of transfer success (input from experimentalist), Ci is the coefficient of colony infection (input from experimentalist).

Taking into consideration the formula (3)

$$P_I = P_Z * L \tag{4}$$

The intensity of model's epizooty is defined as

$$Y = (i_1 + i_2) / B * 100 \tag{5}$$

where Y is the intensity of epizooty, B is the general number of colonies.

The quantity of infected colonies are increasing in the beginning as Fibonacci sequence

$$F_{n+2} = F_n + F_{n+1}, n \in N \tag{6}$$

However when immune colonies appear, the complexity of the epizootic system increases. The description of epizooty by differential equations are not possibe.

The basic parameters described above would be for a case studying the influence of spatial parameters for epizootic processes. It is possible to establish a reverse problem to study the influence of temporary parameter changes at a static, spatial configuration of the settlement. In an ideal case, data can be accumulated on epizootic processes in nature and in experiments on models.

For this study, the plague spread in various settlement types of great gerbils used the following items: 1. The probability of transfer success was equal to one, 2. the coefficient of a colony infection could be varied from one up to 10, and 3. The coefficient of a colony infection was always one.

A two dimensional array of cells was used for the modeling world. The modeling square was equal to 900 ha. The experiment was successful if the epizootic process continued during 18 epochs (model's half year) 11 times without a break. It was a 0.9999 probability that the experiment was nonrandom (Rayfa, 1977). The start value of the coefficient of a colony infection was equal to one. The start density of colonies were 1.33 per ha. The density of colonies declined for 0.11 colonies per ha after each successful experiment for continuous settlements and it was increasing the distance among colonies per one cell on vertical and horizontal lines for narrow-band settlements. As a result, the density of the colonies declined exponentially. If the experiment was not successful the coefficient of a colony infection increase up per unit and recurred again.

3. Results

We gave an example of developmental epizooty in the feigned settlement of the great gerbil (Fig. 5). The goal of the experiment was an examination of qualitative likeness between modeling and natural epizootic processes. The size of the model's working field was 2x2 km. It corresponded to an area of 400 hectares with density of colonies at 2.4 per hectare. A conditional start date of epizooty was March 1st. The epizootic process was developed as a result of the distribution of a plague microbe from one infected and random established colony. The probability of transfer success and the coefficient of a colony infection were constant and equal, 0.8 and 1 respectively.

The epizooty developed slowly at the initial stage. The analysis of the modeled epizootic process showed that the critical time was the initial transfer of a plague microbe between colonies. This was also noted by Soldatkin et al. (1971). The epizootic process could proceed indefinably long if the "infected" colony transferred the plague microbe even to one "not infected" colony during its "infecting" ability. The random hit of a primary "infected" colony on a site of settlement where transfer of a plague microbe would occur raises the probability of epizootic fade-out. The epizooty will end if the transfers of plague microbe do not provide even short-term constant increase of number in the "infected" colonies (1.1 – 2 times during a minimum of 3 cycles of infection). If the start of epizooty was successful, after three months, small foci, consisting of 4 – 8 infected colonies would have been formed (Fig. 5a). The number of the infected colonies increased exponentially. On reaching a certain level of "infected" colonies, the intensity of epizootic process started varying cyclically (Fig. 6). The exponential increase of the infected colonies number was terminated by virtue of two causes. The first cause is the probabilistic character of the plague microbe transfer success from colony to colony. If two or more colonies are infected by an infected colony, the epizootic process intensity will exponentially increase. At the same time, an infected colony can be a source of infection for all neighboring colonies or infect nothing. The second cause is the relatively remote drift of the plague microbe that forms many small foci, consisting of the infected colonies (Fig. 5b). These foci are dissemination centers of epizooty. The spatial structure of the epizooty site is a complicating factor. It begins as a saturated epizooty site of "infected" and "immune" colonies, whereby the plague microbe transfers to other "infected" and "immune" colonies in greater numbers, decreasing the probability of transferring to "non infected" colonies.

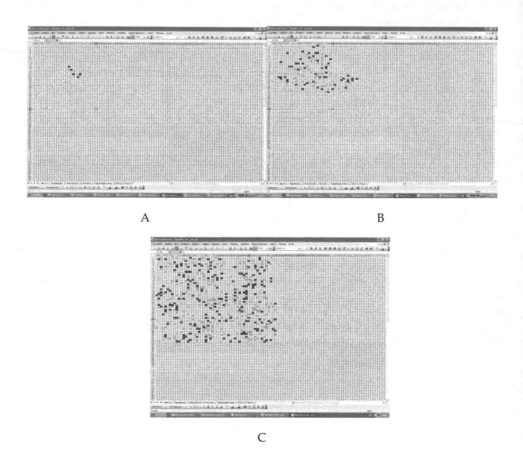

A   B

C

Fig. 5. The visual representation of epizootic process had been developed at model. A) The initial stage of the epizooty development (90 days). B) Remote drift and formation of small foci (210 days). Quantity of the "infected" colonies had started an exponential increase. C) A stage of "saturation" of the epizooty site and transition to cyclic fluctuations

Typically, two peaks of epizooty activity are registered during the year, in the spring and in the fall (21). The epizootic contact rate increases during the rut in early spring and at the end of the spring, as young gerbils explore their territory. Epizootic contact rate and process intensity decreases during the high summer temperatures as the gerbils spend less time on the surface. This also coincides with a generational replacement of the *Xenopsilla* fleas. Contact increases again during the fall when the gerbils become hyperactive, resulting in a second peak of epizooty activity. In autumn, the gerbils cache food and redistribute it among the colonies before winter. The younger generations of fleas are actively feeding, possibly transferring the plague microbe among the animals. This complex mechanism is approximated through the rate of epizooty contact within the model. The exact data about probability of "infecting" colonies in different seasons of the year are absent in literature. Thus, for a qualitative look at the model and nature, the following conclusions were garnered from the model: the probability of a colony infection in the summer is half of that in the spring; a colony infection in the autumn is 0.15% less than in the spring; and a colony infection in the winter is three times below that in the autumn.

Numerical values of the probability of transfer success and the coefficient of colony infections were calculated earlier (Soldatkin et al, 1966, 1971). The probability of transfer success was equal to 0.8 and the coefficient of a colony infection was equal to 1.0, for spring. After entering these parameters, the picture of epizootic process in the model was appreciably changed and looked much like a natural occurrence (Fig. 7).

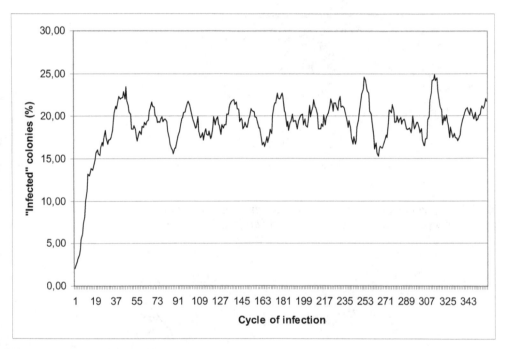

Fig. 6. Intensity of epizootic process in model

There were rare, separate "infected" colonies present in the model's space after the winter period. Those colonies were in the initial stages of the annual cycle of epizooty (Fig. 7a). There were many "immune" colonies that remained after the last epizooty in autumn. The epizooty became sharper at the end of spring when the number of plagued colonies increased, but the number of "immune" colonies decreased (Fig. 7b). In summer, the epizooty recessed and the number of "immune" colonies increased again (Fig. 7c). In autumn, the epizooty raised again as the number of plagued colonies increased and the number of "immune" colonies decreased (Fig. 7d). Thus, the annual epizootic process intensity had qualitative correspondence to a classic, twice-peak epizooty (Fig. 8) as in some natural, plague foci.

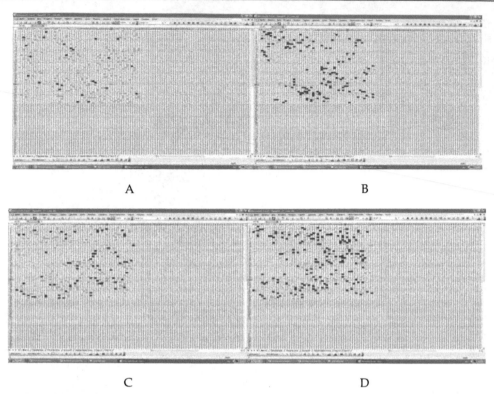

Fig. 7. The epizootic process was developing with a changeable rate of epizootic contact.
Intensity of epizooty in the end: A) winter; B) spring; C) summer; D) autumn

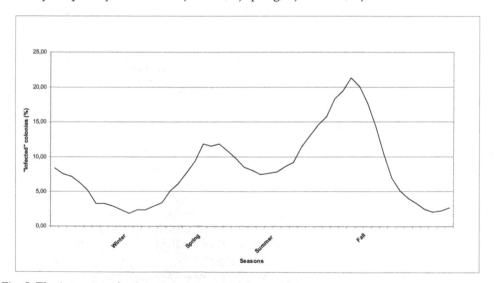

Fig. 8. The intensity of epizootic process modeling within a year

It is interesting, that when a changeable rate of epizootic contact was entered into the model's parameters, the average number of the "infected" and "immune" colonies decreased from 18.92 to 9.31 (p <0.001, Mann-Whitney U test). It was caused by the increase in colonies which were ready for a new infection. As a result, much sharper epizooty could develop. We modeled the sharper epizooty. The coefficient of colony infection was increased to two for this. The picture of epizooty is shown on Fig. 9 and Fig. 10. The epizooty intensity increased almost double in spring and in the fall. However, the number of "infected" colonies was on par with the pre-epizootic level in the winter-summer recession of epizootic activity. The speed of the epizootic spreading fluctuated from 640 to 920 meters per month in these experiments.

It is known that the colonies of the great gerbil have a non-uniform distribution on the surface. According to the distribution pattern, the groups of colonies - the settlement, may be visible as a narrow-band, a wide-band, a large patch, a small patch and continuous settlement (fig 11, 12). Some authors assumed that spatial distribution of the great gerbil settlements played a big role in the epizootic process functioning (Pole, Bibicov, 1991; Dubyanskiy et al., 1992). However that supposition was described as a quality parameter without any quantity values and was quantitatively checked up by modeling. The narrow-band settlements were presented as single colonies, arranged on an identical distance on the vertical and horizontal lines (fig 13).

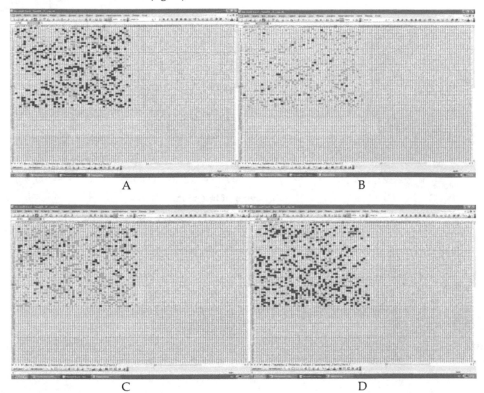

Fig. 9. The epizooty was developing with increasing of the rate of epizootic contact. Intensity эпизоотии in the end A) autumn; B) winter; C) spring; D) summer

First experimental results demonstrated a significant difference of plague spread into continuous and narrow-band settlements (table 1). When the coefficient of a colony infection is one (it is the minimum value of model) the plague can be spread from a density of 1.33 colonies per ha into the continuous settlement. At the same time a density of 0.44 colonies per ha is enough for plague epizootic into the narrow-band settlement. The difference of densities equals three. As the density declines the coefficient of a colony infection proportionally increases into the continuous settlement. A critical threshold is achieved from a density of 0.78 colonies per ha. At the following step of density (0.67 colonies per ha) the epizootic process is recommencing only from value of the coefficient of a colony infection equal 10.

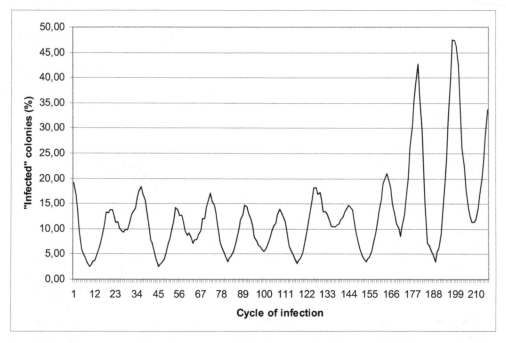

Fig. 10. The intensity of epizootic process which developed at the model with increasing of the rate of epizootic contact (the right part of curve after cycle of infection 171).

In the narrow-band settlement, the minimum limit of density was achieved at 0.11 colonies per ha, and the value of the coefficient of the colony infection was equal to eight. It is not possible to reduce the density of colonies more because of the model limits in the plague microbe drift.

Previous explanation of this phenomenon follows. As monitoring of the model shows, the colonies in the continuous settlement were aggregated to a few groups. The plague epizootic could develop under certain conditions:

1.   if the number of colonies in separate groups are enough for plague circulating only in this group;

2.   if the distance among groups of colonies is not more than a limit of the relatively remote drift;
3.   if the movement activity of animals is enough for connection among the groups of the colonies;
4.   if the rate of epizootic contact is not the highest at the given density of colonies.

These conditions were not often established in the continuous settlement. The plague epizootic was faded by the Allee principle (Odum, 1975). It is possible to check up these items by the statistical data from table 1. The critical threshold of the colonies' density appeared when the average distance among the colonies had exceeded the maximum possibility of the plague drift. At the same time the mode of spacing demonstrated that the circulation plague microbe inside the groups of colonies was not limited by the distance. The rare connections among the colonies and groups of colonies which were situated in the plague microbe drift distance were more stable on the highest level of epizootic contact. It could be about 25% of all colonies at 0.67 colonies per ha density. It demonstrated the first quartile of the distance among colonies. Thus the plague microbe could circulate among 150 colonies in the model's space.

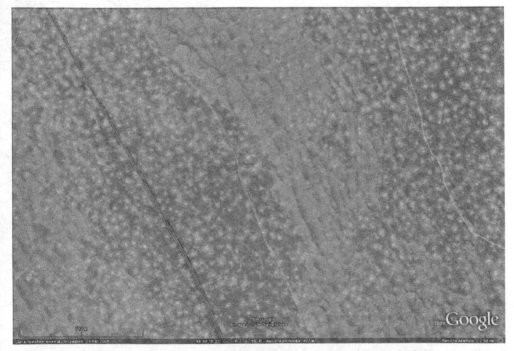

Fig. 11. The continuous settlement of the great gerbil. Colonies of Great Gerbil in a satellite image. Each bright disc represents colonies 10-40m in diameter. The image was captured using the publicly available software Google Earth (http://earth.google.com/). Copyright 2008 DigitalGlobe; Europa Technologies.

Fig. 12. The narrow-band settlement of the great gerbil. Colonies of Great Gerbil in a satellite image. Each bright disc represents colonies 10-40m in diameter. The image was captured using the publicly available software Google Earth (http://earth.google.com/). Copyright 2008 DigitalGlobe; Europa Technologies.

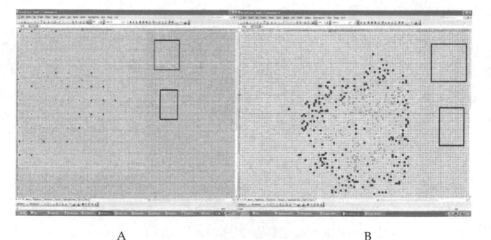

A   B

- plague colony,   - immune colony,   - susceptible colony

Fig. 13. The development of plague epizooty into narrow-band settlement (A) and into continuous settlement (B).

The rate of epizootic contact	Density per ha		The statistical characteristics of distance among colonies (in cells)			
	Continuous settlement	Narrow-band settlement	Average	1 quartile	Mode	Median
1	1.33	0.44	7,33	2	1	5
2	1.11	0.23	8,09	2	0	0
3	1.00	0,17	8,99	2	0	6
4	0.89	0.14	10,11	2	0	7
5	0.78	0,11	10,11	2	0	7
6	0,78	0,11	10,11	2	0	7
7	0,78	0,11	10,11	2	0	7
8	0,78	0.11	10,11	2	0	7
10	0.67	0,11	15,61	4	2	11

Table 1. Dependence of plague spread from colonies density

In the narrow-band settlement the groups of colonies were absent. The colonies were arranged in the form of a lattice and the colonies stood in the vertexes of this lattice. The distance among the colonies was not more than the limit of the relatively remote drift. Each surrounding colony contained enough numbers of colonies for supporting plague epizootic. The colonies were situated in the better places for digging holes in nature (Dubyanskiy, 1970). The migration routes of rodents were not voluntary, and arrangement depended on the settlement structure (Naumov et. al., 1972). The rodents moved along migration routes despite low density, for searching mates, dissemination of young animals, etc., increasing the epizootic contact rate (the dynamic density (Rall, 1965)). The bird-view (fig. 12) demonstrated the limited routes of migration of the animals. These routes are lying along verges of lattice. The similar direction of migration plays a big role in supporting of the plague epizootic.

There is a great difference in the intensity of epizootic processes between the narrow-band and the continuous settlements (fig. 14). The epizooty in the narrow-band settlement is more intensive as in the continuous settlement. The significant difference is represented according to Mann-Whitney U test p<=0.01.

If the epizootic contact rate was variable, the intensity of the epizootic process changed according to the density of the colonies. As the epizootic process was more stable in the narrow-band settlement, this type was used for experiments in the model. When the rate of epizootic contact was to equal one the epizootic process was registered by the minimum experimental density of colonies – 0.11 per ha. However the probability of long-term epizootic process was fewer less - 0.9-0.99.

Is persistence of plague possible when the number of great gerbil is in depression? It is an important question for a problem of an interepizootic plague save solution.

It is known that great gerbils' depression, inhabited colonies are often saved as groups which were situated into better ecological niches (Marin et al., 1970; Burdelov L.A. & Burdelov V.A., 1981; Dubrovskiy et. al., 1989). It was interesting that the number of colonies in separate groups were necessary for persistent plague epizooty. The colonies were arranged as a united group in the model space. The spacing among the colonies and the level (rate) of epizootic contact changed during experiments. Results demonstrated in the table 2.

The minimum number of colonies necessary for plague persistent during the modeling year was equal to 50 with maximum value of the probability of transfer success, the coefficient of the colony infection and the spacing among the colonies. After decreasing the spacing among the colonies to eight cells, the epizootic process was developed inside a group consisting of 40 colonies, but after decreasing the spacing to seven cells, 60 colonies were required. Each decrease in the spacing among the colonies (and increasing of plague contact) the number of colonies necessary for supporting plague epizooty increased. The spatial structure of epizootic process was reminiscent of a "layer cake" (fig. 15) or "wave" of infected colonies. The external layer is a layer of infected coloniesfollowed by a layer of immune colonies.

The "readied" colonies were arranged around the first infected colony inside the "Layer cake". The drift of plague microbe from infected to "readied" colonies was difficult. The irregularity of epizootic process, which is one of the main conditions of long existing epizooty, disappeared. Thus the epizootic process degenerated. If the rate of contact decreases, the effect of "wave" is neutralized. In total the plague can persist in a group of 25 colonies when epizootic contact rate is equal to one. The biology base of this effect was confirmed in nature. When great gerbils' density was in deep depression, the movement of rodents decreased and the level of epizootic contacts decreased too (Naumov et al., 1972).

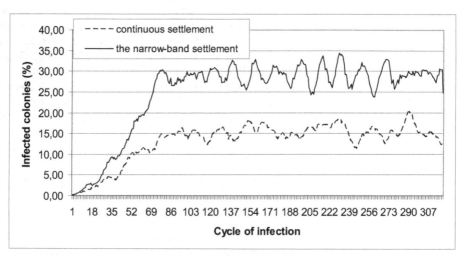

Fig. 14. The intensity of epizootic process in the different type of great gerbils' settlements in model. The density of narrow-band settlement is 0.44 colonies per ha, the density of the continuous settlement is 1.33 colonies per ha.

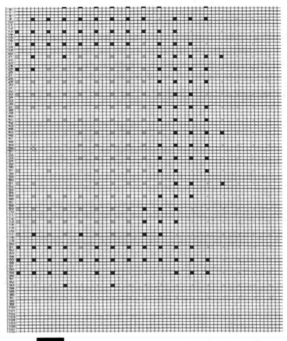

▢ - immune colonies, ■ - infected colonies

Fig. 15. The spatial development of epizooty at the high density of colonies and the highest epizootic contact rate.

The modelling showed that 25 colonies were enough for the plague to persist for a long time. The epizootic contact level had non-line (threshold) dependence from density and number of colonies.

The level of epizootic contact	The colonies quantity	The distance among colonies (cells)
10	50	10
10	40	8
10	60	7
10	50	6
10	70	5
10	400	4
6,48	60	4
1,2	40	2
1,14	30	2
1,0	25	0

Table 2. The level of epizootic contact, the distance among colonies and minimum of colonies quantity, when the plague persistence during 1 year period

4. Discussion

It is important to remember the following for a quality assessment of the modeling and real epizootic process. The investigation of the plague's spatial structure had shown that the small foci of plagued colonies represented pools of the neighbor colonies where there had been living sick or immune gerbils and their infested ectoparasites (Dubyanskiy, 1963; Klassovskiy et al., 1978; Rothschild, 1978; Rivkus et al., 1985). Each pool was irregularly located on the surface and separated by significant distance from other identical pools. It allowed for observing all phases of the epizootic process in a small focus at one time. Sick rodents were observed in some parts of colonies, while an increased fraction of immune rodents were observed in other parts of the colonies. Some colonies were without the plague and these were reserved for epizooty prolongation. All authors agreed on the irregularity of epizootic process in the small focus and permanent moving of epizooty in the territory of the plague focus.

The model-based process was analogous to the described-above epizooty at the small focus or at the pool of the small foci (so-called "epizooty spot"). The analogous types of colonies were present in the model. The well-defined pools of colonies (small foci) were separated from each other and non-uniformity distributed in the working space of the model. The speed of the model-based epizooty spreading coordinated well with field experiments through the marking of animals with radioisotopes and the direct observation of epizooty in nature. The cyclicity arising in the model simulated fluctuations of natural epizootic process intensities (Dubyanskiy and Burdelov, 2010) that confirmed an adequacy of the model-based process to real epizooty.

We have shown the possibility of modeling an annual epizootic cycle by changing the season's epizootic contact rate. This demonstrated the adequacy of the model in transmission theory framework. At the same time, the qualitative picture of epizootic cycle phases was simulated in the model, shown by an increase in the immune colonies' fraction after autumn or spring epizooty peaks. Even a minor increase (up to two) in the single parameter, coefficient of colony infection, was an important factor, as it led to epizooty outbreak. A doubling in the coefficient of colony infection could indicate a gerbil family's contact rate with other families as two times in 10 days instead of only one time. Our thinking, based on literature data, was that both values of the coefficient of colony infection (1 or 2) are usual for the density of a colony, which we used in the model. It demonstrates how minor fluctuations of the epizootic contact rate can lead to the strong changes in the epizootic situation. It is necessary to note that the endless epizootic process can develop even in a small model area. It can be considered as indirect acknowledgement of high survivability of a similar probabilistic process. It was correct for the real epizootic process at natural foci and it was often proved in practice.

It was quantitatively demonstrated in the big role of spatial distribution of great gerbils for plague epizootic. The various kinds of burrow systems constituting settlements, was marked by M. A. Dubyanskiy (1963, 1963a, 1970). He showed that lines of separate kinds of burrow systems played a different role in the plague spread. The modeling demonstrated spatial arrangement is very significant for plague epizootic too. It was known some plague foci had "band" settlements. These are "the foci from North desert subzone". The settlements from these foci are arranged along dry riverbeds, ravines, gullies. The plague in these foci was more intense and had more rare interepizootic periods as in the foci with the "continuous" settlements (Aikimbaev et al., 1987). We believe the spatial kind of great gerbil settlements was a significant influence to this phenomenon.

The background epizooty of low density in the occupancy   colonies have proved the opinion that a negative result of surveillance can result not only from a real absence of plague in the focus, but also from insufficient intensity and extent of the surveillance (Dubyanskiy et al., 2007; Dubyanskiy, 2008). Culling surveillance area of plague foci is possible provided 0.1 – 0.6 colonies per ha.

When the great gerbils' density was set in the deep depression, the plague microbe was not detected. However,   the plague epizooties were restarted after the gerbils' density recovered. Such "disappearing" of the plague microbe, resulted in a long discussion about routes of plague microbe conservation during interepizootic period. It was believed that the plague microbe requires specific mechanisms for conservation as a soil plague and the transmission mechanism by fleas was not effective (Akiev, 1970; Burdelov, 2001; Popov, 2002). Special works were made in Mangyshlak (Marin et al., 1970) in the South Balhash area (Burdelov L.A. & Burdelov V.A., 1981) and in Kizilkum (Dubrovskiy et al., 1989) it was shown that the occupancy colonies were arranged individually as a group, from five colonies to several. It is enough for plague microbes circulating for a long time. At the same time the most difficult to detect plague by unspecific surveillance because the groups of colonies were rare arranged on the biggest (more over 1000 square kilometers) area. Moreover the plague microbe was not conserved for every colony group. It is a cause, possibly, unregistering of epizooty and existing interepizootic periods. The modeling experiments demonstrated that the classical transmission theory can explain a complexity of plague detected during depression of gerbils and a beginnings of plague at a normal density of rodents.

The modeling allows searching numeric parameters of the epizootic process of plague. However, the simulation may appear a perspective method of forecasting plague epizootics, if applied along with permanent comparing and correcting of modeling results with the real surveillance situation.

The number of great gerbil burrow systems was nearest to natural data about quantity and distribution colonies during deep depressions of great gerbil's number. Thus, the permanent plague epizootics could occur even during deep depressions of great gerbils' number.

5. Conclusion

The model can be useful for studying the general appropriateness of the plague's epizootic process and for obtaining formerly inaccessible quantitative characteristics. It is the result of model simplicity, availability of high-performance computing workstations, model parameter input by anyone, and a large display area in the model. The model allows displaying a real and modeled structure of the rodent's settlement. Thus, it was possible to study the spatial regulation of the epizootic process depending on the density of colonies, distribution of colonies, intrasettlement migration, and so forth. Remote sensing methods and GIS can be used as a base for modeling. The basic characteristics of simulated epizootic processes qualitatively also are quantitatively close to natural analogues. It is important to emphasize the model is not constrained by species-specificity. It can be used for modeling epizootic processes among different species and for different vector-borne diseases.

6. References

Addink E.A., S.M. de Jong, S.A. Davis, V. Dubyanskiy, L.A. Burdelov & H. Leirs. (2010). The use of high-resolution remote sensing for plague surveillance in Kazakhstan. *Remote Sensing of Environment* No. 114, pp. 674-681

Akiev A.K. (1970). Condition of a question on studying the mechanism of preservation of the plague in interepizootic years. (Review). *Problems of especially dangerous infections.* Vol 4 (14), pp. 13-33

Aikimbaev A.M., Aubakirov S.A., Burdelov A.S., Klassovskiy L.N., Serjanov O.S. *The Nature desert plague focus of Middle Asia.* Alma-Ata. Nauka. 1987. 207 p.

Alekseev A. F. (1974). The long of live and peculiarity of moving of Great Gerbils in North-West Kizilkum. *The material VIII scientist conference plagues control establishments of Central Asia and Kazakhstan* pp. 220-222, Alma-Ata

Begon M., Klassovskiy N., Ageyev V., Suleimenov B., Atshabar B. et al.. (2006). Epizootiological parameters for plague (Yersinia pestis infection) in a natural reservoir in Kazakhstan. *Emerging Infectious Diseases.* Vol. 12, Issue 2, pp. 268-273

Burdelov L. A. (2001). Sources, the reasons and consequences of crisis in plague epizootology. *Quarantine and zoonotic infections in Kazakhstan,* Vol. 3, pp. 20-25.

Burdelov L. A., Burdelov A. S., Bondar E. P., Zubov V. V., Maslennikova Z. P., Rudenchik N. F. (1984). Hole using of great gerbil (Rhombomys opimus, Rodentia, Cricetidae) and epizootically significance empty colonies in the Central Asia foci. *Zoology journal,* No. LXIII, Vol. 12, pp. 1848-1858.

Burdelov L. A., Burdelov V. A. (1981). *The optimum ecotope and ways of great gerbils survive in unfavorable conditions. The ecology and medical significance gerbils in the USSR fauna, Proceedings of national meeting by ecology and medical significance gerbils as importance rodents of arid zone,* pp 129-130, Moscow, Russia.

Burdelov L. A., Davis S., Dubyanskiy V.M., Ageyev V. S., Pole D. S., Meka-Mechenko V. G., Heier L., Sapojnikov V. I. (2007) The prospect use remote sensing in monitoring of epidemiology of plague. *Quarantine and zoonotic infections in Kazakhstan,* Vol. 1-2 (15-16), pp. 41-48

Davis S., Trapman P., Leirs H., Begon M., Heesterbeek J.A.P. (2008) The abundance threshold for plague as a critical perculation phenomenon. *Nature.* Vol. 454, pp. 634-637

Davis S, Begon M, De Bruyn L, Ageyev VS, Klassovskiy NL, Pole SB, Viljugrein H, Stenseth NC, Leirs H. (2004). Predictive thresholds for plague in Kazakhstan. *Science,* Vol. 304, pp. 736-738. (doi:10.1126/ science.10.95854).

Rivkus J.Z.,Mitropolskiy O.V.,Urmanov R.A.,Belyaeva S.I. (1985). Features of epizootic plague among rodents Kyzylkum, In: *Fauna and ecology of the rodents,* ol. 16. pp. 5-106. Publishing house of Moscow University

Dubrovskiy J. A., Mitropolskiy O.V., Urmanov R.A., Tretjakov G.P., Zakirov R.H., Sarjinskiy B.A., Fotteler E.R. (1989). The natural laws of great gerbil arrangement during perennial depression of its number in Central Kizilkum. *The gerbils - importance rodents of USSR arid zone. Fan, Tashkent,* pp. 42-44.

Dubyanskiy M. A. (1963). About exterior of plague epizooty on a different phases in settlements of great gerbil. *Materials of scientist conference,* pp. 76-78. Alma-Ata

Dubyanskiy M. A. (1970). *The ecology structure of great gerbil's setllements in Priaral Karakum.* Abstract of Ph. D. thesis. Alma-Ata

Dubyanskiy M. A. (1963). The settlement types of great gerbil and theirs epizotology significant in Priaral Karakum *Zoology journal*, Vol. 42, No. 1, pp. 103-113

Dubyanskiy M. A., Kenjebaev A., Stepanov V. V., Asenov G. A., Dubyanskaja L. D. (1992). *Forecasting epizootic activity of a plague in Near- Aral region and Kizilkum.* Nukus: Karakalpakstan.

Dubyanskiy V.M., Burdelov L.A. (2008). A possibility of mathematic modeling of plague epizootic process in GIS-space. «*Modern technologies in the implementation of the global strategy of combating infectious diseases over the territories of the CIS member states*». *Release IX international. science.-pract. conf. CIS member states*, pp. 195-197. Volgograd

Dubyanskiy V.M., Burdelov L.A. (2010). The computer model of plague's epizootic process in Rhombomys opimus settlements and some results of its researching. *Zoology journal*, Vol 89, No 1, pp. 79–87

Dubyanskiy, V. M. (2008). Interval evaluation of intensity of plague epizootics. *Quarantinable and zoonotic infections in Kazakhstan. Almaty*, Vol 1-2 (17-18), pp. 52-57

Dubyanskiy, V. M., Pole S. B., Burdelov L.A., Klassovskaja E.. V., Sapojnikov V. I. (2007). Depending on plague detection in Bakanas plain on completeness of epizootologival survey. *Quarantinable and zoonotic infections in Kazakhstan.* Vol. 1-2 (15-16), pp.62-66.

Dubyanskiy V. M. (2010). Modelling plague spread in the different kind settlements of Great Gerbil (Rhombomys opimus Liht., 1823, Rodentia, Cricetidae), In: *Current issuaes of zoonotic diseases. International conference. 29 March 2010.* Ulaanbaatar, pp. 92-99

Etkins P. (1987). *Order and disorder in Nature.* M.: Mir.

Grabovskiy V. I. (1995). Cellular automata as a simple model of complication system. *Progress of modern biology*, No. 115, pp. 412 – 418.

Gage, K. L., & Kosoy, M. Y. (2005). Natural history of plague: Perspectives from more than a century of research. *Annual Review of Entomology*, No. 50, pp. 505–528

Klassovskij L. N, Kunitskiy V. N., Gauzshtejn D. M, Burdelov A. S., Aikimbaev M. A., Dubovizkiy N. M., Novikov G. S., Rasin B. V., Savelov U.V., (1978). To a question on short-term forecasting epizootic situations in Ili-Raratal country between two rivers. *The condition and prospects prophilaxis of plagues: Thesis on USSR's conference plague control institute "Microbe"*, pp. 69-70. Saratov

Litvin V. U., Karulin B. E., Vodomorin N. A., Ohotskij J. V. (1980). Radioisotope modeling epizootic situations and stochastic model of rabbit-fever epizooty in stack, In: *Fauna and ecolgy of rodents.* Vol. 14, pp. 63-84. Publishing house of Moscow University

Marin S.N., Kamnev P.I., Korinfskiy A.N., Nikitenko G.I. (1970). The depression of great gerbils' number on Mangishlak. *Problems of particularly dangerous infections*, Vol. 4 (14), pp. 149-157.

Naumov N. P., Lobachev V. S., Dmitriev P. P., Smirin V. M. (1972). *The nature plague focus in Priaral Karakum.* Moscow. MGU.

Novikova T. N., Pershin I. B., Gubaydullina G. S., Reshetnikova P. I., Sviridov G. G., Kunitskiy V. N. (1971). Results of serology study of Great Gerbils after its infected by plague fleas. *The material V scientist. conference plague control establishments of Central Asia and Kazakhstan,* pp. 148-151. Alma-Ata

Odum E. (1975). *Fundamentals of ecology.* Moscow. Mir.

Pole S. B., Bibicov D. I.. (1991). The dynamic of structure and the mechanisms supporting the optimal density of grey marmot population. *The structure of marmot's population*, pp. 148-171. Moscow

Popov N.V. (2002). *Discrete character – the basic spatial-temporal peculiarity of plague manifestations in natural foci of souslic type*. Saratov University Press.

Rall J.M. (1965). *Natural nidality and epizootology of plague*. Moscow. Medicina

Rayfa G. (1977). *Decision analysis*. Moscow. Nayka

Rothschild E.V. (1978). *Spatial structure of the natural foci of a plague and methods of its studying*. Moscow. The Moscow State University.

Rothschild E.B., Postnikov G.B., Dankov S.S, Kosarev V.P. (1975). Development of groups of plague holes at epizooty among Great Gerbils. *Problems of particularly dangerous infections*, Vol. 3-4 (43-44), pp. 70-77.

Samsonovich L. G., Peysahis L. A., Gordienko O. J. (1971). Susceptibility, sensitiveness and intensity immunity of Great Gerbils after infected it by leucine-depended strain of plague microbe from different foci. *The material V scientist. conference plague control establishments of Central Asia and Kazakhstan*, pp. 169-172. Alma-Ata

Sedin V.I. (1985). Hole's of Great Gerbils as elements of structure plague epizooty in the Central and the Western Kara-Kum. *XII interrepublican scientist conference plague control establishments of Central Asia and Kazakhstan on prophylaxis of a plague*, pp. 141-143. Alma-Ata

Soldatkin I. S, Rodnikovskiy V. B., Rudenchik J. V. (1973). Experience of statistical modeling of plague epizootic process. *Zoology journal*, Vol. 52, pp. 751-756.

Soldatkin I. S., Rudenchik J. V. (1967). About correlation between model and nature plague epizooty process. *The material V scientist. conference plague control establishments of Central Asia and Kazakhstan, to the 50 anniversary of the Great October Socialist revolution*, pp 70-71. Alma-Ata

Soldatkin I. S., Rudenchik J. V., 1971. Some questions a plague enzooty as a self-regulating system "rodent-flea-causative agent", In: *Fauna and ecology of the rodents*, Vol. 10. pp. 5 – 29. Publishing house of Moscow University

Soldatkin I. S., Rudenchik J. V., Ostrovsky I. B., Levoshina A. I. (1966). Quantitative characteristics of condition development of plague epizooty in Great Gerbil's settlements. *Zoology journal*, Vol. 14, pp. 481-485

10

Rapid Start-Up of the Steam Boiler, Considering the Allowable Rate of Temperature Changes

Jan Taler and Piotr Harchut
Department of Power Installations, Cracow University of Technology
Poland

1. Introduction

The development of wind turbine engineering, which can be characterized by large inconsistencies in the amount of delivered electricity over time, generated new problems regarding the regulation of power engineering systems. At high and low wind speeds, the energy production falls rapidly, meaning that the conventional heat or steam-gas blocks have to be activated very quickly. Steam boilers in blocks have to be designed in a way that allows for a start-up of the block within a few dozen minutes. The main elements limiting the quick activation of steam boilers are thick-walled structural elements, in which heat induced stresses can occur during the start-up cycle.

The start-up and the shut-down processes shall be conducted in a way that ensures that the stresses at the places of stress concentration do not exceed the allowable values.

To be able to ensure the appropriate durability and safety of the blocks adapted for a rapid start up, a precise analysis of the flow-heat processes, together with the strength analysis, is required. Such analysis is covered in this paper.

German boiler regulations, TRD – 301 and the European Standard, EN-12952-3 allow determining two allowable rates for heating the pressure element:

- v_{T1} the pressure p_1 at the beginning of the start-up process,

- v_{T2} at the pressure p_2 at the end of the start-up process.

Heating rates for intermediate pressures shall be determined using the method of linear interpolation (Fig. 1).

To be able to heat the boiler evaporator at the maximum allowable rate, heat flow rate $\dot{Q}$ shall be delivered to the evaporator in the boiler furnace chamber (Fig. 2). The value of $\dot{Q}$ is a function of many parameters, but for the boiler of a specific construction, it depends mainly on the rate of changes of the saturation pressure dp/dt and the mass of the saturated steam mass flow rate.

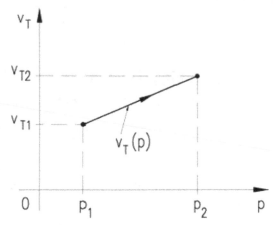

Fig. 1. Heating rate of the pressure boiler element as a pressure function, determined in accordance with TRD-301 regulations.

2. Pressure and temperature changes in the evaporator during the boiler heating process

In accordance with TRD 301 regulations (Fig. 1), the changes of the heating rate $v_\tau = dT_s/dt$ in the pressure function are interpolated using a straight line.

$$\frac{dT_n}{dt} = \frac{p_2 v_{T1} - p_1 v_{T2}}{p_2 - p_1} + \frac{v_{T2} - v_{T1}}{p_2 - p_1} p_n(T_n) \qquad (1)$$

The water saturation pressure $p_s(T_s)$ from Eq. (1) can be expressed in the following way in the temperature function:

$$p_n = \exp(-19{,}710662 + 4{,}2367548 T_n) \qquad (2)$$

Where pressure p_n is expressed in bars and temperature T_n is expressed in °C.

Eq. (1) has been integrated using the Runge-Kutty method of the 4-th rank at the following initial condition:

$$T_n\big|_{t=0} = T_n(p_1) \qquad (3)$$

After determining, by solving the initial condition (1-3), the flow of the saturation temperature in a function of time, that is $T_n(t)$, the pressure changes $p_n(t)$ can be determined, e.g., from function (2). The pressure changes rate dp_s/dt and the saturation temperature changes rate dT_s/dt in a function of time shall be determined on the basis of the assumed run $v_T(p_n)$ (Fig. 1).

In this paper, the dynamics of the boiler evaporator of the OP-210M boiler will be analysed.

OP-210M is a fine coal fuelled boiler with natural circulation. The $T_n(t)$ and $p_n(t)$ flows, determined for $p_1=0$ bar, $v_{T1}=2$ K/min and $p_2=108{,}7$ bar, $v_{T2}=5$ K/min are presented in Fig. 2.

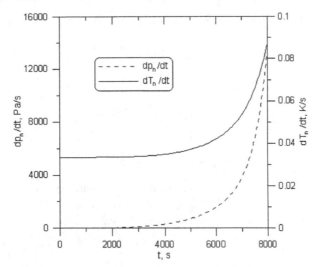

Fig. 2. Temperature and saturation pressure changes in the boiler evaporator during the heating process.

After determining the $T_n(t)$ and dT_n/dt time run, it is possible to determine the rate of pressure changes dp_n/dt using the following formula:

$$\frac{dp_n}{dt} = \frac{dp_n}{dT_n} \frac{dT_n}{dt} \tag{4}$$

Where functions dp_n/dT_n for saturated steam can be determined by the integration of formula (2).

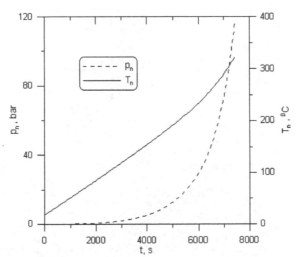

Fig. 3. Rates of the pressure dp_n/dt and temperature dT_n/dt changes, during the evaporator heating process.

The pressure dp_n/dT_n changes rate can also be calculated using an approximate formula:

$$\frac{dp_n}{dt} = \frac{p_n(t+\Delta t) - p_n(t-\Delta t)}{2\Delta t} \tag{5}$$

Where: Δt is the time step.

For the calculations, it can be assumed, for example, that $\Delta t = 1$ s. Temperature dT_n/dt and pressure dp_n/dt changing rates are presented in Fig. 3.

3. Mass and energy balance for the boiler evaporator

Modelling of the transient state effects occurring in the boiler evaporator is usually done while assuming that it is an object having a concentrated mass and heat capacity [4]. The start point for the determination of the heat flow rate $\dot{Q}$, which assures that the evaporator is heated at the desired rate vT(t), are the mass and energy balance equations for the evaporator (Fig. 4):

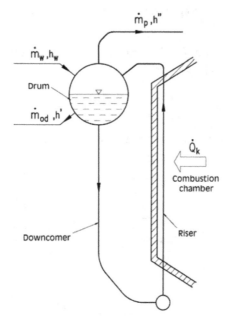

Fig. 4. Diagram of the boiler evaporator.

$$\frac{d(V'\rho' + V''\rho'')}{d\tau} = \dot{m}_w - \dot{m}_p - \dot{m}_{od} \tag{6}$$

$$\frac{d(V'\rho' + V''\rho'')}{d\tau} = \dot{m}_w - \dot{m}_p - \dot{m}_{od} \tag{7}$$

After transforming formulas (6) and (7) we obtain:

$$\dot{Q}_k = \dot{m}_{od} \frac{\rho''}{\rho'-\rho''}(h''-h') - \dot{m}_w \left(h_w - \frac{\rho'h'-\rho''h''}{\rho'-\rho''} \right) + \dot{m}_p \frac{\rho'(h''-h')}{\rho'-\rho''} +$$

$$[V'(\rho' \frac{dh'}{dp} + \frac{\rho''(h''-h')}{\rho'-\rho''} \frac{d\rho'}{dp} - 1) + V''(\rho'' \frac{dh'}{dp} + \frac{\rho(h''-h')}{\rho'-\rho''} \frac{d\rho''}{dp} - 1) + m_m c_m \frac{dT_n}{dp_n}] \frac{dp_n}{dt} \qquad (8)$$

Formula (8) allows determining the heat flow rate $\dot{Q}_i$, which shall be delivered to the evaporator from the furnace chamber in order to ensure the assumed rate of the pressure change dp_n/dt. Changes of the $d\rho'/dp$ and $d\rho''/dp$ functions and the dh'/dp and dh''/dp functions, appearing in formula (8) are presented in Fig. 5 and Fig. 6.

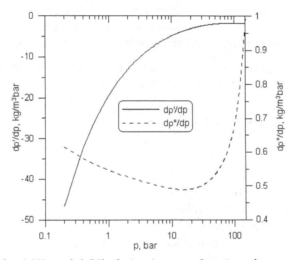

Fig. 5. Changes of the $d\rho'/dp$ and $d\rho''/dp$ derivatives as a function of pressure.

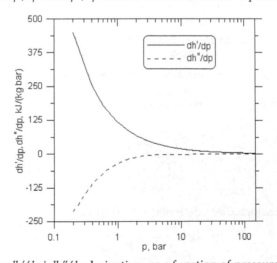

Fig. 6. Changes of the dh'/dp i dh''/dp derivatives as a function of pressure.

The precision in determining those functions is very important. The best way is to determine those functions through analytical differentiation of the $\rho'(p)$ and $\rho''(p)$ functions, obtained earlier using the least square methods on the basis of the pair tables.

4. Heat flow rate absorbed in combustion chamber

From the conducted calculations, we can see, that the heat flow rate $\dot{Q}$ delivered to the evaporator can be lowered significantly by reducing the mass of the steam mass flow rate $\dot{m}_p$ delivered from the boiler drum. The minimum steam mass flow rate $\dot{m}_p$, which shall be produced in the boiler during the start-up process can be determined from the condition, stating that the maximum allowable temperatures for the particular stages of the heater shall not be exceeded. The furnace oil mass flow rate $\dot{m}_{pal}$, delivered to the burners during the boiler start up process, which is necessary to assure the heat flow rate $\dot{Q}$ shall be determined on the basis of the calculations of the boiler furnace chamber.

The heat flow rate $\dot{Q}_{k,kom}$ absorbed by the walls of the boiler furnace chamber, expressed in W, can be calculated from the following formula:

$$\dot{Q}_{k,kom} = \dot{m}_{pal} W_d + \dot{m}_{pow} c_{p,pow} \Big|_0^{Tpow} T_{pow} - \dot{m}_{sp} c_{p,sp} \Big|_0^{T\check{s}p} T''_{sp} \tag{9}$$

$\dot{m}_{pal}$	– fuel mass flow rate, kg/s,
W_d	– fuel calorific value, J/kg,
$\dot{m}_{pow}$	– air mass flow rate, kg/s,
$c_{p,pow}\Big\|_0^{Tpow}$	– average specific heat of air, at constant pressure and at temperature from 0 to T_{pow} [oC], J/(kg·K)
T_{pow}	– temperature of air delivered to the furnace chamber, oC
$\dot{m}_{sp}$	– flue gas mass flow rate, kg/s
$c_{p,sp}\Big\|_0^{T\check{s}p}$	– average specific heat of flue gas, at constant pressure and at temperature from 0 to T''_{sp} [oC], J/(kg·K)
T''_{sp}	– flue gas temperature at the exit of the furnace chamber, oC

The temperature of flue gases at the exit of the furnace chamber, T''_{sp} is calculated from the equation:

$$T''_{sp} = \frac{T_{ad} + 273,15}{M\left(\dfrac{a_p}{Bo}\right)^{0,6} + 1} - 273,15 \tag{10}$$

Where Bo is the Boltzmann number, determined by the formula:

$$Bo = \frac{\dot{m}_{sp} \overline{c}_{p,sp}}{\sigma \psi A_k T_{ad}^3} \tag{11}$$

$\overline{c}_{p,sp}$ is the average specific heat of flue gases in J/(kg·K), at temperature ranging from $T_{\circ}^{*}$ to T_{ad}

$$\overline{c}_{p,sp} = c_{p,sp}\Big|_{T\hat{s}p-273,15}^{Tad-273,15}$$
(12)

a_p – emissivity of the furnace chamber,

M – parameter characterizing the area, in which the flame temperature in the chamber is highest,

A_k – area of the walls of the boiler furnace chamber, m²

T_{ad} – theoretical (adiabatic) combustion temperature, ºC

σ – Stefan-Boltzmann constant, $\sigma = 5{,}67 \cdot 10^{-8}$ W/(m² · K⁴)

ψ – thermal efficiency coefficient of the screens, being the quotient of the heat flux absorbed by the screen and the incident flux.

The calculation of the heat flow rate $\dot{Q}_{k,kom}$ absorbed by the walls of the furnace chamber of the OP-210M boiler has been calculated using (9). In the function of the fuel mass flow rate $\dot{m}_{pal}$ and the air excess coefficient λ, the calorific value of the oil was assumed to be: $W_d = 41060$ kJ/kg.

The results of the calculations are presented in Fig. 7.

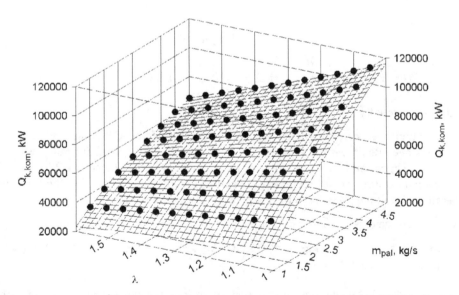

$$z=(a+cx+ey+gx^2+iy^2+kxy)/(1+bx+dy+fx^2+hy^2+jxy)$$
a=-5799.6402 b=0.0491163 c=73693.158 d=0.65706476 e=18923.183 f=-0.00037425411
g=1224.6567 h=-0.55044914 i=-9120.8028 j=0.1281819 k=-34536.155

Fig. 7. Heat flow rate $\dot{Q}_{k,kom}$, in kW, delivered to the evaporator of the OP-210M boiler, as a function of the fuel mass flow rate $\dot{m}_{pal}$ in kg/s and the air excess coefficient λ.

It was proven that the increase of the air excess coefficient λ at the assumed fuel mass flow rate $\dot{m}_{pal}$ leads to the reduction of the heat flux delivered from flue gases in the furnace chamber to the evaporator. This is due to the fact that the temperature of the flue gases in the furnace chamber T_{pl} was lowered significantly, which in turn led to a significant decrease of the heat flow rate $\dot{Q}_{k,kom}$, as the heat flux is proportional to temperature difference between the flame temperature and the temperature of the chamber walls raised to the fourth power, that is $\left(T_{pl}^{4} - T_{sc}^{4}\right)$ when λ increases $\dot{Q}_{k,kom}$ is reduced, and the heat flux absorbed by the steam over heaters increases. Increasing the fuel mass flow rate $\dot{m}_{pal}$ at the assumed air excess coefficient λ leads to the increase of the heat flow rate $\dot{Q}_{k,kom}$ absorbed by the walls of the furnace chamber.

Next, the fuel mass flow rate $\dot{m}_{pal}$ delivered to the boiler during the start-up (heating) process was determined. After determining from (8) the heat flow rate $\dot{Q}_{k}(t)$, the non-linear algebraic equation is solved.

$$\dot{Q}_{k}(t) = \dot{Q}_{k,kom}\left(\dot{m}_{pal}, \lambda\right) \tag{13}$$

$\dot{m}_{p}, \dot{m}_{pal}$ The fuel mass flow rate, at the assumed value of the air excess coefficient λ, is determined from (13). Changes of the mass flow rate in a time function, during the start-up process of the OP-210M boiler, determined from (13), for $\lambda = 1,1$ is presented in Fig. 9a and 9b. The calculations were conducted for different steam mass flows produced in the boiler evaporator. The mass of the fuel oil used for the boiler start up process from time $t=0$ to $t=t_k$ can be calculated using the formula:

$$m_{pal} = \int_{0}^{t_k} \dot{m}_{pal} dt \tag{14}$$

The integral (14) was calculated numerically, using the trapezium rule for integration.

5. Calculation results

The calculation has been completed for the evaporator of the OP-210M boiler, for the pressure change rate dp_n/dt presented in Fig. 3. The temperature of the incoming water T_w during the evaporator heating process is at a given moment lower by 10K than the saturation temperature Tn. The temperature run of the incoming water T_w, the saturation temperature T_n and the pressure in the evaporator p_n are presented in Fig. 8.

The calculations have been conducted for the following data:

$\dot{m}_{ods} = 0,51$ kg/s, $\dot{m}_{w} = 17,1$ kg/s, $\dot{m}_{p} = 16,57$ kg/s, m_m=171900 kg, $c_m = 511$ J/(kg·K), $V' = 43,6$ m³, $V'' = 15,9$ m³.

The results of the $\dot{Q}_{k}$ calculations, determined for various saturated steam mass flow rates, are presented in Fig. 9a and Fig. 9b.

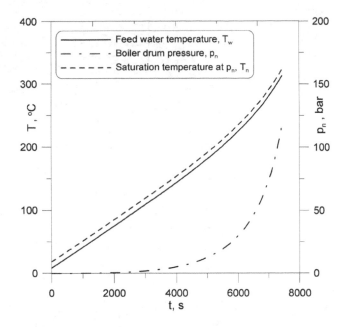

Fig. 8. Temperature run of the incoming water temperature T_w, saturation temperature T_n and pressure in the evaporator p_n during the boiler start up cycle.

After analysing the results shown in Fig. 9a we can see that if the mass capacity of the evaporator $\dot{m}_p$ equals zero, then the flow rate of the heat flow rate $\dot{Q}_t$ delivered to the evaporator shall increase. In this case, all heat delivered from the furnace chamber is used mainly for heating water contained in the boiler evaporator. This is due to the increase of the specific heat of water c_w, which increases in line with the pressure increase. For example, at pressure p_n =1 MPa water specific heat equals: c_w = 4405 J/(kg·K), and at the pressure p_n =10 MPa water specific heat equals: c_w = 6127 J/(kg·K). Moreover, at the end of the start-up process a large increase of the rate of temperature changes in the evaporator was noticed (Fig. 3), which required increased heat flow rate $\dot{Q}_t$ to be delivered to the evaporator.

In larger mass capacities $\dot{m}_p$ the heat flow rate $\dot{Q}_t$, which shall be delivered from the furnace chamber to the evaporator, increases significantly, because there is a large power demand on water evaporation (Fig. 9). For the evaporator efficiencies greater than 90 t/h (Fig. 9b), however, it has been noted that for the assumed flux $\dot{m}_p$, the reduction of the heat flow rate $\dot{Q}_t$ occurred during the end phase of the heating process, despite the increase of the rate of evaporator heating (Fig. 3). This is due to the fact that the water evaporation heat, $r = h'' - h'$ decreases in line with the pressure increase in the evaporator. For example, at the pressure p_n=1 MPa, water evaporation heat equals: r = 2014,4 J/(kg·K), and at the pressure p_n=10 MPa, the water evaporation heat is reduced to: r = 1317,7 J/(kg·K).

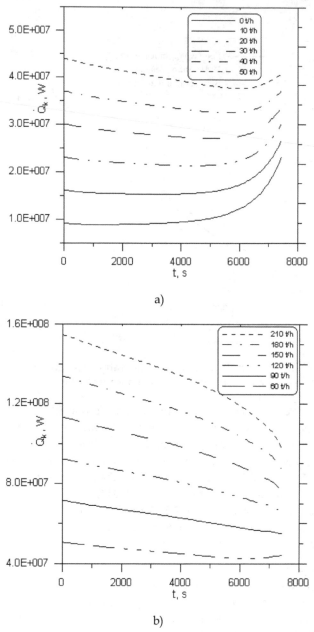

Fig. 9. Heat flow rate $\dot{Q}_k$ delivered to the evaporator of the boiler in the furnace chamber for various steam fluxes $\dot{m}_p$.

Fuel mass flow rate $\dot{m}_{pal}$ during the start-up of the OP-210M boiler obtained from Eq. (13) for the excess air number $\lambda=1,1$ are shown in Figs. 10a and 10b.

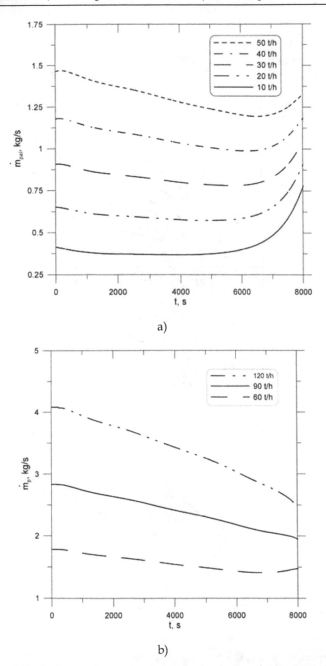

a)

b)

Fig. 10. Changes of the fuel mass flow rate $\dot{m}_{pal}$ in kg/s, during the start-up of the OP-210M boiler, depending on the steam mass flow rate $\dot{m}_p$ in t/h, produced in the boiler evaporator.

The calculations are carried out for various steam flow rates $\dot{m}_p$ generated in the evaporator.

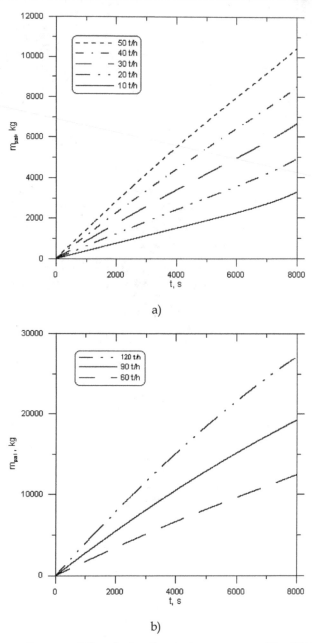

a)

b)

Fig. 11. Fuel consumption m_{pal}, in kg, during the start-up process of the OP-210M boiler, depending on the steam mass flow rate $\dot{m}_p$, in t/h, produced in the boiler evaporator.

The results of the calculations of m_{pal} corresponding to the runs of the fuel mass flow rates $\dot{m}_{pal}$ from Fig. 10 are presented in Fig. 11.

From the analysis of the results shown in Fig. 11 we can see that the fuel consumption depends mainly on the steam mass flow rate $\dot{m}_p$ removed from the boiler drum. During the start-up process, lasting 8000s, the fuel consumption varies from m_{pal} = 3308 kg, at the steam mass flow rate $\dot{m}_p$ = 10 t/h, to m_{pal} = 27207 kg, at the steam mass flow $\dot{m}_p$ = 120 t/h. Due to the high price of the fuel oil, the boiler start up process shall be conducted at the minimum steam mass flow rate $\dot{m}_p$, ensuring the appropriate cooling of the pipes of the over heaters.

Strength analyses show that temperature can change at the beginning of warm-up by leaps and bounds, without exceeding the allowable stress.

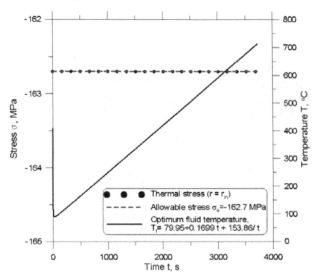

Fig. 12. The allowable process of the temperature due to the thermal stresses on the inner surface of the boiler drum.

The allowable temperature of the process due to the thermal stresses on the inner surface of the drum can be approximated by using:

$$T_n(t) = a + bt + \frac{c}{t^2} \tag{15}$$

where the constants a, b and c are determined from the condition:

$$\sigma_\varphi(r_{in}, t_i) \cong \sigma_a \quad t_i = (i-1)\Delta t \quad i = 1, 2, 3, ..., n \tag{16}$$

Condition (16) means that for optimal temperature of the process defined by formula (15) constants a, b and c should be chosen so that the thermal stress on the inner circumferential surface of the drum was, for n time points, equal to the allowable stress. In the case of the drum analysed in this paper, for allowable stress $\sigma_a = -162,7$ MPa, the following constants were obtained: a=79,95°C, b=0,1699 °C/s, c=153,86 s. Starting with an optimal start-up boiler water temperature change in the drum is very difficult, because the optimal temperature is very high at the beginning of start-up and then very quickly reduced to the minimum value.

Therefore, during start-up an approximation of optimal temperature change will be implemented by the step up temperature increase at the beginning and then heating at a constant speed. When the pressure in the drum is higher than the atmospheric pressure, after the jump of temperature, the speed of temperature change is equal v_{T1} (at the beginning) and v_{T2} (at the end of start-up). Determination of the heating or cooling rate of the drum for any pressure p is the same as in TRD 301 regulations or European Standard BS EN 12952-3.

Fig. 13a and fig. 13b show the changes in temperature and pressure of water in the drum at different initial temperature jumps.

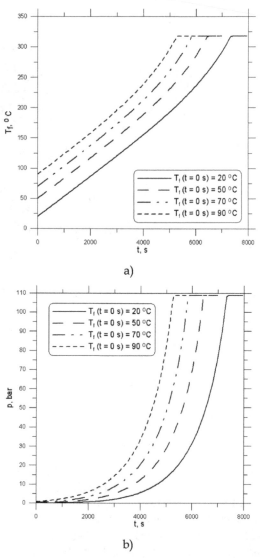

a)

b)

Fig. 13. Temperature and pressure of the water in the drum during boiler start-up.

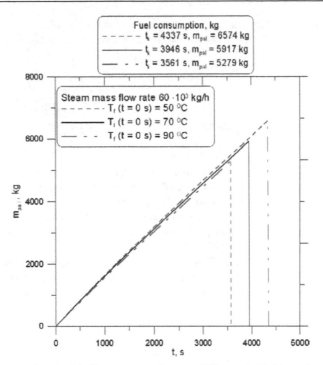

Fig. 14. Fuel consumption and boiler start-up time at different initial jumping changes in water temperature in the evaporator boiler.

An analysis of the results in Fig. 14 shows that the release of the initial jump in the temperature of the boiler start-up reduces the time and reduces fuel consumption.

6. Conclusions

The model of the dynamics of the boiler evaporator, presented in this paper, can be used for the determination of the heat flux, flowing from flue gases in the furnace chamber to the boiler evaporator, necessary to warm up the boiler at the assumed rate of the medium temperature change in the boiler evaporator. The fuel mass flow rate $\dot{m}_{pal}$, for which the heat flux delivered to the evaporator equals $\dot{Q}_k$, is determined on the basis of the calculations of the boiler furnace chamber. Fuel consumption during the boiler start up process, necessary for heating the boiler evaporator to the assumed temperature, from the assumed nominal start temperature, assuming that the heating of the boiler drum is conducted at the maximum allowable speed, due to the limitations caused by heat stresses at the boiler drum/down comers intersection, has been determined. In order to limit the consumption of the expensive fuel oil during the boiler start up process, the process shall be conducted at the minimum steam mass flow rate, ensuring the appropriate cooling of the over heaters and at the minimum allowable air coefficient. The dependencies derived in this paper can be used for the preparation of the optimal technology of the boiler start up process.

It is shown that allowing the rapid increase of water temperature at the start of commissioning of the boiler reduces the boiler start-up time and helps to reduce fuel oil consumption.

7. References

TRD 301 (2001), Zylinderschalen unter innerem Überdruck. *Technische Regeln für Dampfkessel* (TRD), pp. 143-185, Heymanns Beuth Köln – Berlin, Germany.

Staff Report, (2007). Dealing with cycling: TRD 301 and the Euro Norm compared, *Modern Power Systems*, Vol 27, No 5, p 33-38.

European Standard, EN 12952-3, *Water-tube boilers and auxiliary installations – Part 3: Design and calculation for pressure parts* , CEN – European Committee for Standardization, rue de Stassart 36, B-1050 Brussels, 25. July 2001.

Profos, P., (1962). *Die Regelung von Dampfanlagen,* Springer, Berlin, Germany.

Styrykowicz, M. A., Katkowskaja, K. J., Sierow, E. P. (1959). *Kotielnyje agregaty.* Gosenergoizdat, Moscow, Russia.

Sierow, E.P., Korolkow, B. P. (1981). *Dinamika parogienieratorow.* Energija, Moscow, Russia.

Taler, J., Dzierwa, P. (2010): A new method for optimum heating of steam boiler pressure components. *International Journal of Energy Research*, Volume 35, Issue 10, pages 897–908, Wiley, August 2011.

11

Optical Interference Signal Processing in Precision Dimension Measurement

Haijiang Hu[1,*], Fengdeng Zhang[2], Juju Hu[1] and Yinghua Ji[1]

[1]Jiangxi Normal University, Nanchang,
[2]University of Shanghai for Science and Technology, Shanghai,
China

1. Introduction

The optical techniques are widely used in the automatic measurement and test. In the precision dimension measurement, there are some optical instruments that are widely applied for the microscale or nanoscale dimension measurement, such as the grating, the homodyne interferometer, and the heterodyne interferometer. An optical instrument system is often divided into two main subsystems: the optical subsystem and the electrical subsystem. Moreover, some optical instrument systems also have the mechanical subsystem and the computer subsystems. The optical subsystem includes optical source and other optical elements, such as the polarizing beam splitter, the quarter-wave plate, and so on. The electronic subsystem turns the received optical signals into the electrical signals by the photoelectric receiver, and uses the analog signal system and the digital signal system to process the received signals. A typical structure of the optical instruments is shown as Fig.1.

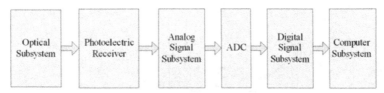

Fig. 1. A typical structure of optical instruments: ADC, analog-digital converter.

In the performance evaluation of the optical measurement instrument, the resolution and the precision (JCGM, 2008) are two important parameters. The resolution describes the ability of a measurement system to resolve detail in the object that is measured. The precision of a measurement system, also called reproducibility or repeatability, is the degree to show the same results when the measurement process is repeated under the unchanged conditions. This chapter mainly discusses two topics: the signal processing method for the improvement of measurement resolution and the elimination errors for the improvement of measurement precision. Generally speaking, the optical source, the design of optical path, and the signal processing are the main factors to improve the measurement resolution. For

* Corresponding Author

example, in a heterodyne Michelson interferometer with double path optical difference that is shown as Fig.2, the traveling distance of the micro-displacement platform L is as follows:

$$L = \frac{\lambda}{2} \int_0^T \Delta f dt, \tag{1}$$

where λ is the wavelength of laser, and Δf is the Doppler frequency shift. Obviously, λ is smaller, the measurement resolution is higher.

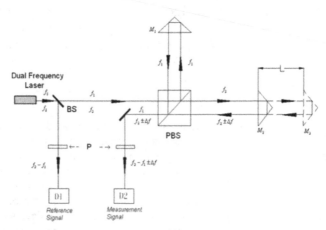

Fig. 2. A heterodyne Michelson interferometer with double path optical difference: BS, beam splitter; PBS, polarizing beam splitter; P, polarization analyzer.

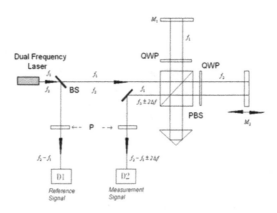

Fig. 3. A heterodyne Michelson interferometer with fourfold path optical difference: BS, beam splitter; PBS, polarizing beam splitter; P, polarization analyzer; QWP, quarter-wave plate.

The multiple reflection of optical path, also called the optical subdivision, is the second factor to influence the measurement resolution. A heterodyne Michelson interferometer with fourfold path optical difference is shown as Fig.3. When the micro-displacement platform M_2 moves the same distance L in Fig.2 and Fig.3, the Doppler frequency shift Δf in Fig.3 is twice as large as Δf in Fig.2. In the same configuration of laser source and electrical signal processing system, the measurement resolution in Fig.3 is twice as large as the measurement resolution in Fig.2. Therefore, the optical subdivision relies on the optical path design of optical system and can be realized by the multiple reflections in the optical path (Cheng et al., 2006).

The third method is the electrical signal processing. Because of the light attenuation in optical parts, the optical subdivision is hard to realize the high subdivision number. Moreover, the measurement error from the nonideal characters and arrangement of optical parts is gradually increased when the number of optical subdivision is added. So the electrical signal processing becomes the most useful method to improve the measurement resolution.

Moreover, the detection and elimination of errors is the important research targets for the improvement of measurement precision. These errors have very serious influence on the measurement precision. In this chapter we discuss electrical signal processing of orthogonal signals (fringes), which is widely used in the optical gratings, the homodyne interferometers, and other optical instruments.

2. Signal preprocessing

In the optical gratings and the homodyne interferometers, the photoelectric converter turns the intensity of light into the electrical signal. Because of the requirement of the direction recognition when the measured object moves, the orthogonal signals whose phase difference is $\pi/2$ are used in these instruments. In order to obtain the orthogonal signals, 4-channel signal receiving systems are designed in the optical systems. Fig.4 shows a typical 4-channel signal receiving systems (Keem et al., 2004).

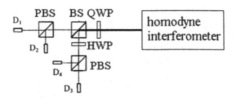

Fig. 4. 4-channel signal receiving systems: BS, beam splitter; PBS, polarizing beam splitter; QWP: quarter-wave plate; HWP: half-wave plate;

When the measured object moves forward, the 4-channel electrical signals are collected from the photosensors and shown as follows:

$$\begin{cases} I_1 = A_1 + B_1 \sin\theta \\ I_2 = A_2 - B_2 \cos\theta \\ I_3 = A_3 - B_3 \sin\theta \\ I_4 = A_4 + B_4 \cos\theta \end{cases} \quad (2)$$

where A_n is the DC component of channel n, B_n is the amplitude of AC component of channel n, and θ is the change of phase that is related to the displacement ΔL when the measured object moves, respectively. In the homodyne interferometer with two path optical differences, the relationship between θ and ΔL is shown as follows:

$$\theta = \frac{4\pi}{\lambda}\Delta L ,\tag{3}$$

where λ is the wavelength of laser.

Two signals I_x and I_y can be obtained by subtracting:

$$\begin{cases} I_y = I_1 - I_3 \\ I_x = I_4 - I_2 \end{cases}.\tag{4}$$

From Eqs.2 and Eqs. 4,

$$\begin{cases} I_y = A_1 - A_3 + (B_1 + B_3)\sin\theta \\ I_x = A_4 - A_2 + (B_2 + B_4)\cos\theta \end{cases}.\tag{5}$$

Under the ideal condition, $A_2=A_4$, $A_1=A_3$ and $B_2+B_4 = B_1+B_3$. So, two orthogonal signals are obtained as follows when the measured object moves forward:

$$\begin{cases} I_y = B\sin\theta \\ I_x = B\cos\theta \end{cases},\tag{6}$$

where $B=B_1+B_3=B_2+B_4$.

Similarly, the 4-channel signals are shown as follows when the measured object moves backward:

$$\begin{cases} I_1 = A_1 + B_1\sin\theta \\ I_2 = A_2 + B_2\cos\theta \\ I_3 = A_3 - B_3\sin\theta \\ I_4 = A_4 - B_4\cos\theta \end{cases},\tag{7}$$

where the parameters of Eqs.7 are the same as the parameters of Eqs.2. Similarly, from Eqs.4 and Eqs.7, two orthogonal signals are obtained as follows when the measured object moves backward:

$$\begin{cases} I_y = B\sin\theta \\ I_x = -B\cos\theta \end{cases},\tag{8}$$

where $B=B_1+B_3=B_2+B_4$.

So, ΔL is obtained by using the half periodic counting of orthogonal signals and the calculation of θ as follows:

$$\Delta L = \frac{\lambda}{2}N + \varepsilon \,, \tag{9}$$

with

$$\varepsilon = \frac{\Delta\theta}{4\pi}\lambda \,,$$

where λ is the wavelength of laser, N is the counting value of half period, and ε is the additional displacement value that is shorter than $\lambda/2$ and can be calculated from the orthogonal signals, respectively. For the improvement of measurement resolution, some bidirectional subdivision methods are studied to raise the resolution of N and ε.

3. Bidirectional subdivision

3.1 λ/4 bidirectional subdivision

The simplest bidirectional subdivision method (Yoshizawa, 2005) is shown as Fig.5. This method is applied on both the analog orthogonal signals and the digital orthogonal signals. The principle of this method is shown in the right part in Fig.5. During the measured object moves forward, when signal u_2 crosses zero from positive to negative, signal u_1 is positive; when signal u_1 crosses zero from positive to negative, signal u_2 is negative; when signal u_2 crosses zero from negative to positive, signal u_1 is negative; when signal u_1 crosses zero from negative to positive, signal u_{2f} is positive. Therefore, the counting of four times in a period for forward movement is realized.

Similarly, during the measured object moves backward, when signal u_2 crosses zero from negative to positive, signal u_1 is positive; when signal u_1 crosses zero from positive to negative, signal u_2 is positive; when signal u_2 crosses zero from positive to negative, signal u_1 is negative; when signal u_1 crosses zero from negative to positive, signal u_2 is negative. From the above detection, the counting of four times in a period for bidirectional movement is realized.

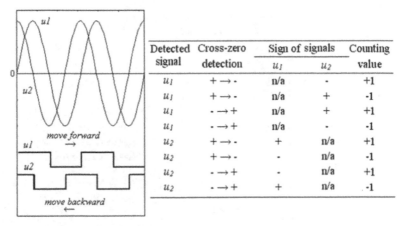

Detected signal	Cross-zero detection	Sign of signals u_1	u_2	Counting value
u_1	$+ \rightarrow \cdot$	n/a	$\cdot$	+1
u_1	$+ \rightarrow \cdot$	n/a	$+$	-1
u_1	$\cdot \rightarrow +$	n/a	$+$	+1
u_1	$\cdot \rightarrow +$	n/a	$\cdot$	-1
u_2	$+ \rightarrow \cdot$	$+$	n/a	+1
u_2	$+ \rightarrow \cdot$	$\cdot$	n/a	-1
u_2	$\cdot \rightarrow +$	$\cdot$	n/a	+1
u_2	$\cdot \rightarrow +$	$+$	n/a	-1

Fig. 5. λ/4 bidirectional subdivision

3.2 λ/8 bidirectional subdivision

In order to realize the counting of more than four times in a period, some methods are studied. The first method uses two thresholds to realize λ/8 bidirectional subdivision by (Chu et al., 2003). The method defines two thresholds T_h and T_l. T_h is the amplitude of normalized signal in the $\pi/4$ phase while T_l is the negative of T_h. When u_1 cross across T_h from the bottom to the top, the movement direction is positive if u_2 is positive at the same time or negative if u_2 is negative at the same time. Similarly, when u_1 cross across T_l from the bottom to the top, the movement direction is positive if u_2 is positive at the same time or negative if u_2 is negative at the same time. By combination between λ/4 bidirectional subdivision and two thresholds detection, the λ/8 bidirectional subdivision is realized.

(Cui et al., 2000) proposes a method to realize λ/8 bidirectional subdivision by the constructed function. The constructed function is as follows:

$$f = |u_1| - |u_2|. \tag{10}$$

Fig.6 shows the function images of u_1, u_2 and f. According to the change of sign of u_1, u_2 and f, λ/8 bidirectional subdivision is achieved. The principle of this method is as shown in Fig.7.

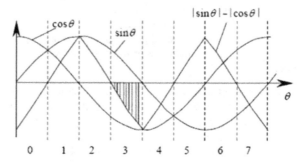

Fig. 6. Function images of u₁, u₂ and f

sinθ	cosθ	f	code	status	
> 0	> 0	< 0	110	A	*move forward*
> 0	> 0	> 0	000	B	A->B->C->D->
> 0	< 0	> 0	101	C	E->F->G->H->A
> 0	< 0	< 0	100	D	
< 0	< 0	< 0	000	E	*move backward*
< 0	< 0	> 0	001	F	A->H->G->F->
< 0	> 0	> 0	011	G	E->D->C->B->A
< 0	> 0	< 0	010	H	

Fig. 7. Principle of λ/8 bidirectional subdivision by Eq.10

3.3 $\lambda/16$ bidirectional subdivision

Similar to (Cui et al., 2000), (Chen et al., 2005) proposes another constructed function and status code method to achieve $\lambda/16$ bidirectional subdivision. The difference from (Cui et al., 2000), (Chen et al., 2005) uses 4 bits code as status parameter.

3.4 $\lambda/2^{n+2}$ bidirectional subdivision

(Hu et al., 2009) and (Hu & Zhang, 2012) proposes a method of $\lambda/2^{n+2}$ bidirectional subdivision that uses both the constructed function and the cross-zero detection. The $\lambda/8$ and $\lambda/16$ bidirectional subdivision methods are simply introduced at first. Commonly, two orthogonal signals are defined as follows when the measured object moves forward:

$$\begin{cases} u_{1f} = U_0 \sin\theta \\ u_{2f} = U_0 \cos\theta \end{cases}' \tag{11}$$

where U_0 is the amplitude of the orthogonal signals, and θ is the change of phase that is related to the displacement when the measured object moves.

The following two relative functions are constructed to be the reference signals in order to achieve the bidirectional subdivision:

$$\begin{cases} r_{11f} = U_0 \sin\theta + U_0 \cos\theta \\ r_{12f} = U_0 \sin\theta - U_0 \cos\theta \end{cases}. \tag{12}$$

Reference signals r_{11f} and r_{12f} can be acquired by the digital adder. Therefore, the $\lambda/8$ subdivision of the two orthogonal signals can be realized easily by the zero-cross detection which consists of the orthogonal signals and the reference signals.

Based on the $\lambda/4$ bidirectional subdivision When the positive voltage drives the measured object to move forward, reference signal r_{12f} crosses zero from negative to positive, signals u_{1f} and u_{2f} are positive; when reference signal r_{11f} crosses zero from positive to negative, signal u_{1f} is positive and signal u_{2f} is negative; when reference signal r_{12f} crosses zero from positive to negative, signals u_{1f} and u_{2f} are negative; when reference signal r_{11f} crosses zero from negative to positive; signal u_{1f} is negative and signal u_{2f} is positive; Therefore, the counting of eight times in a period for the forward movement is realized.

The same to the forward counting, it's easy to realize the counting of eight times in a period for the backward movement. When the measured object moves backward, the two orthogonal signals are defined as follows:

$$\begin{cases} u_{1b} = U_0 \sin\theta \\ u_{2b} = -U_0 \cos\theta \end{cases}. \tag{13}$$

The same to Eqs.12, the two relative functions are constructed to be the reference signals:

$$\begin{cases} r_{11b} = U_0 \sin\theta - U_0 \cos\theta \\ r_{12b} = U_0 \sin\theta + U_0 \cos\theta \end{cases}. \tag{14}$$

Similarly, the counting of eight times in a period for the backward movement is also achieved. So, a set of reference signals are only built as follows:

$$\begin{cases} r_{11} = u_1 + u_2 \\ r_{12} = u_1 - u_2 \end{cases}.$$ (15)

The bidirectional subdivision of $\lambda/8$ can be realized by the zero-cross detection of the orthogonal signals and the reference signals.

On the basis of the $\lambda/8$ subdivision, a group of the reference signals is built which is constituted by the two sets of reference signals. During the time of forward counting, the reference signal sets are defined as follows:

$$\begin{cases} r_{11f} = U_0 \sin\theta + U_0 \cos\theta \\ r_{12f} = U_0 \sin\theta - U_0 \cos\theta \end{cases},$$ (16)

$$\begin{cases} r_{21f} = U_0 \sin 2\theta + U_0 \cos 2\theta \\ r_{22f} = U_0 \sin 2\theta - U_0 \cos 2\theta \end{cases}.$$ (17)

In order to calculate the reference signals r_{21f} and r_{22f}, the two orthogonal signals are used to express them. r_{21f} is obtained as follows:

$$r_{21f} = U_0 \sin 2\theta + U_0 \cos 2\theta = U_0[2\cos\theta(\sin\theta + \cos\theta) - 1].$$ (18)

Similarly, r_{22f} is obtained as follows:

$$r_{22f} = U_0[2\sin\theta(\sin\theta + \cos\theta) - 1].$$ (19)

$U_0=1$ is still assumed. The function images of the two orthogonal signals and reference signals for the $\lambda/16$ subdivision are shown in Fig. 8 and Fig. 9 during the time of the forward counting and the backward counting. The same as the $\lambda/8$ subdivision, the cross-zero detection is also used to realize the bidirectional subdivision of $\lambda/16$. Therefore, on the basis of the $\lambda/8$ subdivision, the bidirectional subdivision of $\lambda/16$ for the orthogonal signals is realized by adding a set of reference signals:

$$\begin{cases} r_{21} = \dfrac{2u_2}{U_0}(u_1 + u_2) - U_0 \\ r_{22} = \dfrac{2u_1}{U_0}(u_1 + u_2) - U_0 \end{cases}.$$ (20)

Based on the $\lambda/8$ subdivision and the $\lambda/16$ subdivision, n sets of reference signals can be built to realize the $\lambda/2^{n+2}$ bidirectional subdivision as follows:

$$\begin{cases} r_{n1} = U_0 \sin 2^{n-1}\theta + U_0 \cos 2^{n-1}\theta \\ r_{n2} = U_0 \sin 2^{n-1}\theta - U_0 \cos 2^{n-1}\theta \end{cases},$$ (21)

where n is integer and $n>0$.

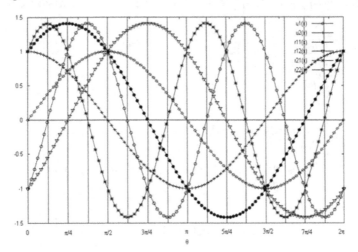

Fig. 8. Forward orthogonal signals and reference signals for $\lambda/16$ subdivision

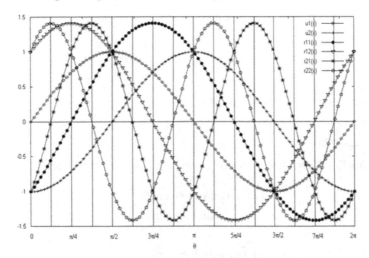

Fig. 9. Backward orthogonal signals and reference signals for $\lambda/16$ subdivision

From Euler's formula:

$$e^{j\theta} = \cos\theta + j\sin\theta,$$

(22)

the following equation is obtained:

$$e^{jn\theta} = \cos n\theta + j\sin n\theta = (\cos\theta + j\sin\theta)^n.$$

(23)

Then Eq. 23 is expanded as follows:

$$\cos n\theta + j\sin n\theta = \binom{n}{0}\cos^n\theta + j\binom{n}{1}\cos^{n-1}\theta\sin\theta + j^2\binom{n}{2}\cos^{n-2}\theta\sin^2\theta + \ldots + j^n\binom{n}{n}\sin^n\theta$$

$$=\begin{cases}\binom{n}{0}\cos^n\theta + j\binom{n}{1}\cos^{n-1}\theta\sin\theta - \ldots + \binom{n}{n}\sin^n\theta, n = 4N \\[2mm] \binom{n}{0}\cos^n\theta + j\binom{n}{1}\cos^{n-1}\theta\sin\theta - \ldots + j\binom{n}{n}\sin^n\theta, n = 4N+1 \\[2mm] \binom{n}{0}\cos^n\theta + j\binom{n}{1}\cos^{n-1}\theta\sin\theta - \ldots - \binom{n}{n}\sin^n\theta, n = 4N+2 \\[2mm] \binom{n}{0}\cos^n\theta + j\binom{n}{1}\cos^{n-1}\theta\sin\theta - \ldots - j\binom{n}{n}\sin^n\theta, n = 4N+3\end{cases} \qquad (24)$$

where N is integer and $N \geq 0$.

From Eqs.24, the following equations are obtained:

$$\begin{cases}\sin 2\theta = 2\sin\theta\cos\theta \\ \cos 2\theta = \cos^2\theta - \sin^2\theta\end{cases} \qquad (25)$$

$$\begin{cases}\sin n\theta = \binom{n}{1}\cos^{n-1}\theta\sin\theta - \binom{n}{3}\cos^{n-3}\theta\sin^3\theta + \ldots - \binom{n}{n-1}\cos\theta\sin^{n-1}\theta \\[2mm] \cos n\theta = \cos^n\theta - \binom{n}{2}\cos^{n-2}\theta\sin^2\theta + \ldots + \sin^n\theta\end{cases}, n = 4N, \qquad (26)$$

where N is integer and $N \geq 1$.

So, from Eqs.16, Eqs.25 and Eqs.26, the following equations are obtained:

$$\begin{cases}r_{11} = U_0\sin\theta + U_0\cos\theta \\ r_{12} = U_0\sin\theta - U_0\cos\theta\end{cases}, n = 1, \qquad (27)$$

$$\begin{cases}r_{21} = U_0\left[2\cos\theta(\sin\theta + \cos\theta) - 1\right] \\ r_{22} = U_0\left[2\sin\theta(\sin\theta + \cos\theta) - 1\right]\end{cases}, n = 2, \qquad (28)$$

$$\begin{cases}r_{n1} = U_0\left(\cos^{2^{n-1}}\theta + \binom{2^{n-1}}{1}\cos^{2^{n-1}-1}\theta\sin\theta - \binom{2^{n-1}}{2}\cos^{2^{n-1}-2}\theta\sin^2\theta - \ldots + \sin^{2^{n-1}}\theta\right) \\[3mm] r_{n2} = U_0\left(-\cos^{2^{n-1}}\theta + \binom{2^{n-1}}{1}\cos^{2^{n-1}-1}\theta\sin\theta + \binom{2^{n-1}}{2}\cos^{2^{n-1}-2}\theta\sin^2\theta - \ldots - \sin^{2^{n-1}}\theta\right)\end{cases}, n > 2 \quad (29)$$

where n is integer.

The same as the $\lambda/8$ subdivision and the $\lambda/16$ subdivision, u_1 and u_2 are used to express r_{n1} and r_{n2}:

$$\begin{cases} r_{11} = u_1 + u_2 \\ r_{12} = u_1 - u_2 \end{cases}, n = 1, \tag{30}$$

$$\begin{cases} r_{21} = \dfrac{2u_2}{U_0}(u_1 + u_2) - U_0 \\ r_{22} = \dfrac{2u_1}{U_0}(u_1 + u_2) - U_0 \end{cases}, n = 2, \tag{31}$$

$$\begin{cases} r_{n1} = \dfrac{1}{U_0^{2^{n-1}-1}}\left(u_2^{2^{n-1}} + \binom{2^{n-1}}{1} u_2^{2^{n-1}-1} u_1 - \binom{2^{n-1}}{2} u_2^{2^{n-1}-2} u_1^2 - \ldots + u_1^{2^{n-1}} \right) \\ r_{n2} = \dfrac{1}{U_0^{2^{n-1}-1}}\left(-u_2^{2^{n-1}} + \binom{2^{n-1}}{1} u_2^{2^{n-1}-1} u_1 + \binom{2^{n-1}}{2} u_2^{2^{n-1}-2} u_1^2 - \ldots - u_1^{2^{n-1}} \right) \end{cases}, n > 2, \tag{32}$$

where n is integer.

For $n=1$, the number of subdivision times is 8.

For $n=2$, the number of subdivision times is 16. This is just shown in Fig.9 during the time of the forward counting and Fig.10 during the time of the backward counting.

Similarly, the $\lambda/32$, $\lambda/64$, $\lambda/128$, ..., $\lambda/2^{n+2}$ subdivision can be obtained where n is integer and $n>2$.

Consequently, a conclusion is obtained as follows:

On the basis of the $\lambda/2^{n+1}$ bidirectional subdivision, the $\lambda/2^{n+2}$ bidirectional subdivision of the orthogonal signals can be realized by adding the two reference signals r_{n1} and r_{n2} and their zero-cross detection. The zero-cross detection uses the original orthogonal signals and the reference signals r_{11}, r_{12}, r_{21}, r_{22}, ..., $r_{(n-1)1}$ and $r_{(n-1)2}$ as the positive-negative judgment items for the forward counting and the backward counting.

4. Error detection and elimination

In the ideal conditions, the phase difference of the orthogonal signals is $\pi/2$. However, because of the imperfect design and manufacture of measurement system, the environmental disturbance and the system noise, the signals often have some errors, such as the nonorthogonality, the non-equality of amplitude, the drift of DC signals, and so on (Heydemann, 1981), which have very serious influence on the precision of the fringe subdivision. The typical orthogonal signals with errors are shown as follows:

$$\begin{cases} u_1^d = u_1 + p \\ u_2^d = \dfrac{1}{r}(u_2 \cos a - u_1 \sin a) + q \end{cases}, \tag{33}$$

where p, q are drift of DC, r is non-equality of amplitude coefficient, a is nonorthogonal error, respectively.

(Heydemann, 1981) proposes a universal method to compensate the errors. From Eqs.33, another equation is obtained as follows:

$$(u_1 + p)^2 + \left(\frac{1}{r} u_2 \cos a - \frac{1}{r} u_1 \sin a + q \right)^2 = U_0^2 , \tag{34}$$

and

$$(u_1^d - p)^2 + \left[\frac{(u_2^d - q)r + (u_1^d - p)\sin a}{\cos a} \right]^2 = U_0^2 . \tag{35}$$

From Eq.35, the error factors are obtained as follows:

$$A u_1^{d2} + B u_2^{d2} + C u_1^d u_2^d + D u_1^d + E u_2^d = 1 , \tag{36}$$

where

$$A = (R^2 \cos^2 a - p^2 - r^2 q^2 - 2rpq \sin a)^{-1}$$
$$B = Ar^2$$
$$C = 2Ar \sin a \qquad \qquad ,$$
$$D = -2A(p + rq \sin a)$$
$$E = -2Ar(rq + p \sin a)$$

respectively. Obviously this is an elliptic function express. The parameter A, B, C, D, and E are called the error factors. Therefore, if the error factors A, B, C, D, and E are obtained, p, q, r, a are calculated as follows:

$$\begin{cases} a = \arcsin C (4AB)^{-\frac{1}{2}} \\ r = (\frac{B}{A})^{\frac{1}{2}} \\ p = \dfrac{2BD - EC}{C^2 - 4AB} \\ q = \dfrac{2AE - DC}{C^2 - 4AB} \end{cases} \tag{37}$$

From Eqs.37, u_1 and u_2 are obtained as follows:

$$\begin{cases} u_1 = u_1^d - p \\ u_2 = \dfrac{1}{\cos a} \left[(u_1^d - p)\sin a + r(u_2^d - q) \right] \end{cases} \tag{38}$$

(Heydemann, 1981) sets up a complete method for the error compensation. This method needs at least a period of signals for the calculation of the error factors. Therefore it has the poor real-time performance. (Zumberge et al., 2004, Eom et al., 2001) also calculate error factors for the compensation by least-squares fitting. (Song et al., 2000) uses phase-modulated grating to compensate for nonorthogonal errors. (Keem et al., 2004) uses Jones matrix calculation to investigate the remaining error of a homodyne interferometer with a quadrature detector system. (Cheng et al., 2009) uses the normalized waveforms to eliminate DC offsets and amplitude variation. (Hu & Zhang, 2011) proposes a method for detection and elimination of the nonorthogonal errors based on digital sampled signals as follows when (Hu et al., 2009, Hu & Zhang, 2012) are applied in the fringe subdivision.

In (Birch, 1990), when the measured object moves forward, the two orthogonal signals which only contain nonorthogonality error are shown as follows:

$$\begin{cases} u_1 = U_0 \sin\theta \\ u_{2f} = U_0 \sin(\theta + \dfrac{\pi}{2} + \Delta\varphi) \end{cases},$$ (39)

where $\Delta\varphi$ is the nonorthogonality error. Similarly, when the measured object moves backward, the original signals which only contain the nonorthogonality error are shown as follows:

$$\begin{cases} u_1 = U_0 \sin\theta \\ u_{2b} = U_0 \sin(\theta - \dfrac{\pi}{2} - \Delta\varphi) \end{cases}.$$ (40)

The detection method of $\Delta\varphi$ is as follows:

At first, a counter is defined as C that records the sample number of the original signal u_2. Then, in the forward movement of the measured object, when the original signal u_1 crosses zero from negative to positive, the counter C resets itself and begins to count. When the original signal u_2 crosses zero, the value of the counter C_e is saved, and the counter C goes on counting. When the original signal u_1 crosses zero from positive to negative, the value of the counter C_{hp} is saved. Consequently, the nonorthogonality error of two original signals can be calculated as follows:

$$\Delta\varphi = \frac{\frac{1}{2}C_{hp} - C_e}{C_{hp}}\pi .$$ (41)

Similarly, when the measured object moves backward, the counter C resets itself and begins to count. When the original signal u_2 crosses zero, the value of the counter C_{eb} is saved and the counter C goes on counting. When the original signal u_1 crosses zero from positive to negative, the value of the counter C_{hp} is saved. the nonorthogonality error of two original signals is calculated as follows:

$$\Delta\varphi = \frac{C_{eb} - \frac{1}{2}C_{hp}}{C_{hp}}\pi .$$ (42)

The precision of $\Delta\varphi$ is related with the sample number in the half period obviously. When the sample number in the half period is higher, the precision of $\Delta\varphi$ is higher. Moreover, the multiple-averaging technique can be used based on sampling many times to improve the computational accuracy.

If d is defined as the displacement value when the original signal moves a period, the relationship between the displacement L and the phase θ is as follows:

$$L = Nd + \frac{\theta}{2\pi}d, \theta \in [0, 2\pi), \qquad (43)$$

where N is the number of periods when the measured object moves, which can be obtained by the cross-zero detection, and θ can be acquired by the method of subdivision and the counting. If the nonorthogonality error does not exist, Eq. 43 can be represented by the $\lambda/8$ bidirectional subdivision from (Hu et al., 2009) as follows:

$$L = \frac{N_8 d}{8} + \frac{4\theta}{\pi}d, \theta \in [0, \frac{\pi}{4}), \qquad (44)$$

where N_8 is the number of $1/8$ periods when the counting system uses $\lambda/8$ subdivision algorithm, and

$$N = \frac{N_8}{8}. \qquad (45)$$

When the nonorthogonality error exists, the algorithm can not divide a signal period into eight subintervals evenly. Therefore Eq.44 is useless if the nonorthogonality error exists. The relationship of the displacement L and θ has to be redefined.

C_p is defined as the sampling value of the signal in a period and can be obtained by means of the cross-zero detection and the counting for the original signals. When the nonorthogonality error does not exist, Eqs.43 and 44 can be shown as follows:

$$L = Nd + \frac{n}{C_p}d, n \in \left(-C_p, C_p\right), \qquad (46)$$

and

$$L = \frac{N_8 d}{8} + \frac{n}{C_p}d, n \in \left(-\frac{C_p}{8}, \frac{C_p}{8}\right), \qquad (47)$$

where n is positive when the measured object moves forward or negative when the measured object moves backward. In the Eq. 46, n is the count value of the original signal when the original signal does not pass through a complete period. Similarly, in the Eq.47, n is the count value of the original signal when the original signal does not pass through one eighth periods.

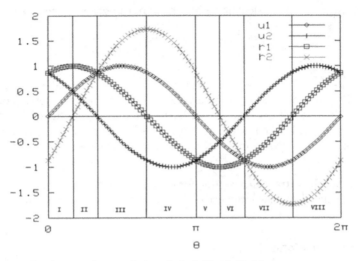

Fig. 10. Subintervals of nonorthogonal signals in $\lambda/8$ subdivision

When the nonorthogonality error exists, eight subintervals can be arranged in two groups. It is shown as Fig. 12. The first group includes the subinterval I, II, V, and VI, while the second group includes the subinterval III, IV, VII, and VIII. Every subinterval in a same group has the same interval length. So the interval length of eight subintervals can be obtained as Table 1:

Number of subinterval	Range of θ in $[0,2\pi)$	Interval length
I	$[0, \dfrac{\pi}{4} - \dfrac{\Delta\varphi}{2})$	$\dfrac{\pi}{4} - \dfrac{\Delta\varphi}{2}$
II	$[\dfrac{\pi}{4} - \dfrac{\Delta\varphi}{2}, \dfrac{\pi}{2} - \Delta\varphi)$	$\dfrac{\pi}{4} - \dfrac{\Delta\varphi}{2}$
III	$[\dfrac{\pi}{2} - \Delta\varphi, \dfrac{3\pi}{4} - \dfrac{\Delta\varphi}{2})$	$\dfrac{\pi}{4} + \dfrac{\Delta\varphi}{2}$
IV	$[\dfrac{3\pi}{4} - \dfrac{\Delta\varphi}{2}, \pi)$	$\dfrac{\pi}{4} + \dfrac{\Delta\varphi}{2}$
V	$[\pi, \dfrac{5\pi}{4} - \dfrac{\Delta\varphi}{2})$	$\dfrac{\pi}{4} - \dfrac{\Delta\varphi}{2}$
VI	$[\dfrac{5\pi}{4} - \dfrac{\Delta\varphi}{2}, \dfrac{3\pi}{2} - \Delta\varphi)$	$\dfrac{\pi}{4} - \dfrac{\Delta\varphi}{2}$
VII	$[\dfrac{3\pi}{2} - \Delta\varphi, \dfrac{7\pi}{4} - \dfrac{\Delta\varphi}{2})$	$\dfrac{\pi}{4} + \dfrac{\Delta\varphi}{2}$
VIII	$[\dfrac{7\pi}{4} - \dfrac{\Delta\varphi}{2}, 2\pi)$	$\dfrac{\pi}{4} + \dfrac{\Delta\varphi}{2}$

Table 1. The interval length of eight subintervals

So the displacement L can be calculated when the nonorthogonality error exists:

$$L = (\frac{\pi - 2\Delta\varphi}{8\pi}N_{8a} + \frac{\pi + 2\Delta\varphi}{8\pi}N_{8b})d + \frac{n}{C_p}d \,, \qquad (48)$$

where N_{8a} is the count value of the subinterval when the original signal passes by the subinterval I, II, V, and VI, N_{8b} is the count value of the subinterval when the original signal passes by the subinterval III, IV, VII, and VIII, and n is the count value of the original signal when the original signal does not pass through a subinterval. N_{8a}, N_{8b}, and n are positive when the measurement platform moves forward or negative when the measurement platform moves backward.

Therefore, without the extra compensation for the nonorthogonality error, the displacement L can be also calculated and the influence of the nonorthogonality error can be eliminated by means of Eq.48.

5. Implementation platform

With the development of electronic and computer technologies, many platforms are widely used in the signal processing for orthogonal signals, such as the hardware (Birch, 1990, Downs & Birch, 1983), FPGA (Hu et al., 2009), the software programming (Zhang et al., 1994, Su et al., 2000), LabVIEW (Yacoot et al., 2001), DSP(Zumberge et al., 2004), and so on. Besides the analog signal processing and the digital signal processing, the image processing is also an important method to realize the fringe analysis (Takeda et al., 1982, Qian, 2004). The image processing system often get data from the CMOS or CCD cameras, and use digital image processing methods to calculate the phase shift. The image processing technique offers many new solutions for the fringe analysis (Patorski & Styk, 2006, Marengo-Rodriguez et al., 2007, Bernini et al., 2009, Larkin et al., 2001, Pokorski & Patorski, 2010, Zhang, 2011) and extends the application fields of fringe analysis (Geng, 2011, Riphagen et al., 2008).

6. Summary

In this chapter the signal processing methods for orthogonal signals are introduced. The orthogonal signals are widely used in the optical grating and the homodyne interferometers. In order to improve the resolution of these optical instruments, some bidirectional subdivision methods are introduced. Besides the subdivision methods, the algorithms for error detection and elimination are also stated for improving the measurement precision. With the development of the hardware technology and the software technology, the precision dimension measurement technology will be further developed and improved on the resolution, the precision, the real-time characteristics, the applicability, and the maneuverability.

7. References

JCGM. (2008). International vocabulary of metrology. Available from:
        <http://www.bipm.org/utils/common/documents/jcgm/JCGM_200_2008.pdf>.

Cheng, Z. Gao, H. Zhang, Z. Huang, H. & Zhu, J. (2006). Study of a dual-frequency laser interferometer with unique optical subdivision techniques. *Appl. Opt.*, Vol. 45, pp. 2246-2250.

Keem, T. Gonda, S. Misumi, I. Huang, Q. & Kurosawa T. (2004). Removing Nonlinearity of a Homodyne Interferometer by Adjusting the Gains of its Quadrature Detector Systems. *Appl. Opt.*, vol. 43, pp. 2443-2448.

Yoshizawa, T. (2008). *Handbook of Optical Metrology - Principles and Applications*, CRC Press.

Chu, X. Lü, H. & Cao, J. (2003). Research on direction recognizing and subdividing method for Moiré(interference) fringes. *Chin. Opt. Lett.*, vol. 1, pp. 692-694.

Cui, J. Li, H. & Chen, Q. (2000). New digital subdividing snd rester-sensing technique for moiré fringe of grating. *Opt. Tech.*, vol. 26, pp. 591-593.

Chen, B. Luo, J. & Li, D. (2005). Code counting of optical fringes: methodology and realization. *Appl. Opt.*, vol. 44, pp. 217-222.

Hu, H. Qiu, X. Wang, J. Ju, A. & Zhang, Y. (2009). Subdivision and direction recognition of $\lambda/16$ of orthogonal fringes for nanometric measurement. *Appl. Opt.*, vol. 48, pp. 6479-6484.

Hu, H. & Zhang, F. (2012). Advanced bidirectional subdivision algorithm for orthogonal fringes in precision optical measurement instruments. *Optik*, in press. Available from <http://www.sciencedirect.com/science/article/pii/S0030402611005973>.

Heydemann, P. L. M. (1981). Determination and correction of quadrature fringe measurement errors in interferometers. *Appl. Opt.*, vol. 20, pp. 3382-3384.

Zumberge, M. A. Berger, J. Dzieciuch, M. A. & Parker R. L. (2004). Resolving quadrature dringes in real time. *Appl. Opt.*, vol. 43, pp. 771-775.

Eom, T. B. Kim, J. Y. & Jeong, K. (2001). The dynamic compensation of nonlinearity in a homodyne laser interferometer. *Meas. Sci. and Technol.*, vol. 12, pp. 1734-1738.

Song, J. H. Kim, K. C. & Kim, S. H. (2000). Reducing tilt errors in moiré linear encoders using phase-modulated grating. *Rev. Sci. Instrum.*, vol. 71, pp. 2296-2300.

Cheng, F. Fei, Y. T. & Fan, G. Z. (2009). New method on real-time signal correction and subdivision for grating-based nanometrology. *Proc. SPIE.*, vol. 7284, 728403.

Birch, K. P. (1990). Optical fringe subdivision with nanometric accuracy. *Precis. Eng.*, vol. 12, pp.195-198.

Downs, M. J. & Birch, K. P. (1983). Bi-directional fringe counting interference refractometer. *Precis. Eng.*, vol. 5, pp. 105-110.

Zhang, G. X. Wang, C. H. & Li, Z. (1994). Improving the accuracy of angle measurement system with optical grating. *CIRP Ann. Manuf. Technol.*, vol. 43, pp. 457-460.

Su, S. Lü, H. Zhou, W. & Wang, G. (2000). A software solution to counting and subdivision of Moiré fringe with wide dynamic range. *Proc. SPIE*, vol.4222, pp. 308-312.

Yacoot, A. Kuetgens, U. Koenders, L. & Weimann, T. (2001). A combined scanning tunnelling microscope and x-ray interferometer. *Meas. Sci. Technol.*, vol. 12, pp. 1660.

Takeda, M. Ina, H. & Kobayashi, S. (1982). Fourier-transform method of fringe-pattern analysis for computer-based topography and interferometry. *J. Opt. Soc. Am.*, vol. 72, pp. 156-160.

Qian, K. (2004).Windowed Fourier Transform for Fringe Pattern Analysis. *Appl. Opt.*, vol. 43, pp. 2695-2702.

Patorski, K & Styk, A. (2006). Interferogram intensity modulation calculations using temporal phase shifting: error analysis. *Opt. Eng.*, vol. 45, 085602.

Marengo-Rodriguez, F. A. Federico, A. & Kaufmann, G. H. (2007). Phase measurement improvement in temporal speckle pattern interferometry using empirical mode decomposition. *Opt. Commun.*, vol. 275, pp. 38-41.

Bernini, M. B. Federico, A. & Kaufmann, G. H. (2009). Normalization of fringe patterns using the bidimensional empirical mode decomposition and the Hilbert transform. *Appl. Opt.*, vol. 48, pp. 6862-6869.

Larkin, K. G. Bone, D. J. & Oldfield, M. A. (2001). Natural demodulation of two-dimensional fringe patterns. I. General background of the spiral phase quadrature transform. *J. Opt. Soc. Am.*, vol. 18, pp.1862–1870.

Pokorski, K. & Patorski, K. (2010). Visualization of additive-type moiré and time-average fringe patterns using the continuous wavelet transform. *Appl. Opt.*, vol. 49, pp. 3640–3651.

Zhang, S. (2011). High-resolution 3-D profilometry with binary phase-shifting methods. *Appl. Opt.*, 50(12), 1753-1757.

Geng, J. (2011). Structured-light 3D surface imaging: a tutorial. *Adv. Opt. Photon.*, vol. 3, pp. 128-160.

Riphagen, J. M. van Neck, J. W. van Adrichem, & L. N. A. (2008). 3D surface imaging in medicine: a review of working principles and implications for imaging the unsedated child. *J. Craniofac. Surg.*, pp. 19, vol. 517-524.

Hu, H. & Zhang, F. (2011). Real-time detection and elimination of nonorthogonality error in interference fringe processing. *Appl. Opt.*, vol. 50, pp. 2140-2144.

Permissions

The contributors of this book come from diverse backgrounds, making this book a truly international effort. This book will bring forth new frontiers with its revolutionizing research information and detailed analysis of the nascent developments around the world.

We would like to thank Dr. Florian Kongoli, for lending his expertise to make the book truly unique. He has played a crucial role in the development of this book. Without his invaluable contribution this book wouldn't have been possible. He has made vital efforts to compile up to date information on the varied aspects of this subject to make this book a valuable addition to the collection of many professionals and students.

This book was conceptualized with the vision of imparting up-to-date information and advanced data in this field. To ensure the same, a matchless editorial board was set up. Every individual on the board went through rigorous rounds of assessment to prove their worth. After which they invested a large part of their time researching and compiling the most relevant data for our readers. Conferences and sessions were held from time to time between the editorial board and the contributing authors to present the data in the most comprehensible form. The editorial team has worked tirelessly to provide valuable and valid information to help people across the globe.

Every chapter published in this book has been scrutinized by our experts. Their significance has been extensively debated. The topics covered herein carry significant findings which will fuel the growth of the discipline. They may even be implemented as practical applications or may be referred to as a beginning point for another development. Chapters in this book were first published by InTech; hereby published with permission under the Creative Commons Attribution License or equivalent.

The editorial board has been involved in producing this book since its inception. They have spent rigorous hours researching and exploring the diverse topics which have resulted in the successful publishing of this book. They have passed on their knowledge of decades through this book. To expedite this challenging task, the publisher supported the team at every step. A small team of assistant editors was also appointed to further simplify the editing procedure and attain best results for the readers.

Our editorial team has been hand-picked from every corner of the world. Their multi-ethnicity adds dynamic inputs to the discussions which result in innovative outcomes. These outcomes are then further discussed with the researchers and contributors who give their valuable feedback and opinion regarding the same. The feedback is then collaborated with the researches and they are edited in a comprehensive manner to aid the understanding of the subject.

Apart from the editorial board, the designing team has also invested a significant amount of their time in understanding the subject and creating the most relevant covers. They scrutinized every image to scout for the most suitable representation of the subject and create an appropriate cover for the book.

The publishing team has been involved in this book since its early stages. They were actively engaged in every process, be it collecting the data, connecting with the contributors or procuring relevant information. The team has been an ardent support to the editorial, designing and production team. Their endless efforts to recruit the best for this project, has resulted in the accomplishment of this book. They are a veteran in the field of academics and their pool of knowledge is as vast as their experience in printing. Their expertise and guidance has proved useful at every step. Their uncompromising quality standards have made this book an exceptional effort. Their encouragement from time to time has been an inspiration for everyone.

The publisher and the editorial board hope that this book will prove to be a valuable piece of knowledge for researchers, students, practitioners and scholars across the globe.

List of Contributors

David Oluwashola Adeniji
University of Ibadan, Nigeria

Gleison Baiôco, Anilton Salles Garcia and Giancarlo Guizzardi
Federal University of Espírito Santo, Brazil

Stefan Brachmanski
Wroclaw University of Technology, Poland

Daniel Bernardon, Vinícius Garcia and Luciano Pfitscher
Federal University of Santa Maria, Brazil

Mauricio Sperandio and Wagner Reck
Federal University of Pampa, Brazil

Antonio Chialastri
Medicair, Rome, Italy

Zheng Liu
Otto-von-Guericke University, Magdeburg, Germany

Nico Suchold and Christian Diedrich
Institut für Automation und Kommunikation (ifak), Magdeburg, Germany

Edward Chikuni
Cape Peninsula University of Technology, South Africa

Josep Ferrer and Marta Pena
Universitat Politecnica de Catalunya, Spain

Carmen Ortiz-Caraballo
Universidad de Extremadura, Spain

Vladimir Dubyanskiy
Stavropol Plague Control Research Institute, Russian Federation

Leonid Burdelov
M. Aikimbaev's Kazakh Science Center of Quarantine and Zoonotic Diseases, Republic of Kazakhstan

J. L. Barkley
Virginia Polytechnic Institute and State University, USA

Jan Taler and Piotr Harchut
Department of Power Installations, Cracow University of Technology, Poland

Haijiang Hu, Juju Hu and Yinghua Ji
Jiangxi Normal University, Nanchang, China

Fengdeng Zhang
University of Shanghai for Science and Technology, Shanghai, China

Printed in the USA
CPSIA information can be obtained
at www.ICGtesting.com
JSHW011427221024
72173JS00004B/704